Thomas Jue • Kazumi Masuda
Editors

Application of Near Infrared Spectroscopy in Biomedicine

Volume 4

Editors
Thomas Jue
Biochemistry and Molecular Medicine
University of California Davis
Davis, CA, USA

Kazumi Masuda
Kanazawa University
Kanazawa, Japan

ISBN 978-1-4614-6251-4 ISBN 978-1-4614-6252-1 (eBook)
DOI 10.1007/978-1-4614-6252-1
Springer New York Heidelberg Dordrecht London

Library of Congress Control Number: 2012956439

Printed on acid-free paper

Springer is part of Springer Science+Business Media (www.springer.com)

Handbook of Modern Biophysics

Series Editor
Thomas Jue
University of California Davis
Davis, California

For further volumes:
http://www.springer.com/series/7845

Preface

The advent of near-infrared spectroscopy (NIRS) presents a unique tool for understanding the regulation of oxidative metabolism during the transition from rest to an active state. Many laboratories have started to apply NIRS to interrogate both cerebral and muscle metabolism and have garnered insights to discriminate the bioenergetics and hemodynamics of healthy and diseased tissue. Yet using NIRS technology and methodology appropriately requires a solid understanding of the principles of physics, biochemistry, and physiology. Indeed, introducing a complex biophysics topic in an academically rigorous but interesting way often poses a challenge.

In keeping with the style of the *Handbook of Modern Biophysics*, the current volume balances the need for physical science/mathematics formalism with a demand for biomedical perspectives. Each chapter divides the presentation into two major parts: the first establishes the conceptual framework and describes the instrumentation or technique, while the second illustrates current applications in addressing complex biology questions. With the additional sections on further reading, problems, and references, the interested reader can explore some chapter ideas more widely.

In the fourth volume in this series, *Application of Near-Infrared Spectroscopy in Biomedicine*, the authors have laid down a solid biophysical foundation. Masatsugu Niwayama and Yutaka Yamashita open by delineating the different types of NIRS methods, describing different instrumentations, and explaining the underlying idea about photon migration. Eiji Okada expands on the key concept of photon migration, especially as it applies to brain imaging. Hajime Miura surveys the application of NIRS in the clinic, while Takafumi Hamaoka describes the use of NIRS in studying human locomotion. Kazumi Masuda explores the use of NIRS to understand regulation of intracellular and vascular oxygen from the start of muscle contraction. Williams and Ponganis show the unique application of NIRS to investigate oxygen regulation in marine mammals during a breathhold or a dive. Finally, Chung and Jue compare the use of NIRS and NMR in determining the role of intracellular oxygen during muscle contraction.

This volume continues the philosophy behind the *Handbook of Modern Biophysics* in providing the reader with a fundamental grasp of concepts and applications on current biophysics topics.

Davis, CA, USA　　　　Thomas Jue

Contents

Contributors

Youngran Chung Biochemistry and Molecular Medicine, University of California Davis, Davis, CA 95616-8635, USA

Miura Hajime Laboratory for Applied Physiology, University of Tokushima, Tokushima 770-8502, Japan

Takafumi Hamaoka Department of Sport and Health Science, Ritsumeikan University, 1 Nojihigashi, Kusatsu, Shiga, Japan

Thomas Jue Biochemistry and Molecular Medicine, University of California Davis, Davis, CA 95616-8635, USA

Kazumi Masuda Faculty of Human Sciences, Kanazawa University, Kanazawa, Ishikawa 920-1192, Japan

Masatsugu Niwayama Department of Electrical and Electronic Engineering, Shizuoka University, 3-5-1 Johoku, Nakaku, Hamamatsu, Shizuoka 432-8561, Japan

Eiji Okada Department of Electronics and Electrical Engineering, Keio University, Yokohama 223-8522, Japan

Paul J. Ponganis Center for Marine Biomedicine and Biotechnology, Scripps Institution of Oceanography, University of California San Diego, La Jolla, CA 92093-0204, USA

Cassondra L. Williams Department of Ecology and Evolutionary Biology, University of California Irvine, 321 Steinhaus, Irvine, CA 92697-2525, USA

Yutaka Yamashita Central Research Laboratory, Hamamatsu Photonics KK, 5000 Hirakuchi, Hamakita-ku, Hamamatsu City, Shizuoka 434-8601, Japan

Principles and Instrumentation 1

Yutaka Yamashita and Masatsugu Niwayama

1.1 Light Absorption and Light Scattering

Quantification of chromophore concentration is based on the Beer-Lambert law [1–3]. In Fig. 1.1 the absorption coefficient μ_a is defined as

$$dI = -\mu_a\, I\, dl, \tag{1.1}$$

where dI is the change in intensity I of light moving along an infinitesimal path dl in a homogeneous medium. Integration over a thickness l (mm) yields

$$I_1 = I_0 \exp(-\mu_a\, l), \tag{1.2}$$

where I_0 is the incident light intensity. This equation is also expressed as base 10 logarithms as

$$I_l = I_0\ 10^{-c\varepsilon l}. \tag{1.3}$$

Where c is the concentration of the compound, and ε is the molar absorption coefficient.

Transmission T is defined as the ratio of the transmitted light intensity to the incident light intensity, I_1/I_0. The optical density (OD) is given by

$$\mathrm{OD} = \log 10(1/T) = \log 10(I_0/I_1). \tag{1.4}$$

Y. Yamashita, BS
Central Research Laboratory, Hamamatsu Photonics KK, 5000 Hirakuchi,
Hamakita-ku, Hamamatsu City,
Shizuoka 434-8601, Japan
e-mail: yutaka@crl.hpk.co.jp

M. Niwayama, Ph.D. (✉)
Department of Electrical and Electronic Engineering, Shizuoka University,
3-5-1 Johoku, Nakaku, Hamamatsu, Shizuoka 432-8561, Japan
e-mail: tmniway@ipc.shizuoka.ac.jp

T. Jue and K. Masuda (eds.), *Application of Near Infrared Spectroscopy in Biomedicine*,
Handbook of Modern Biophysics 4, DOI 10.1007/978-1-4614-6252-1_1,

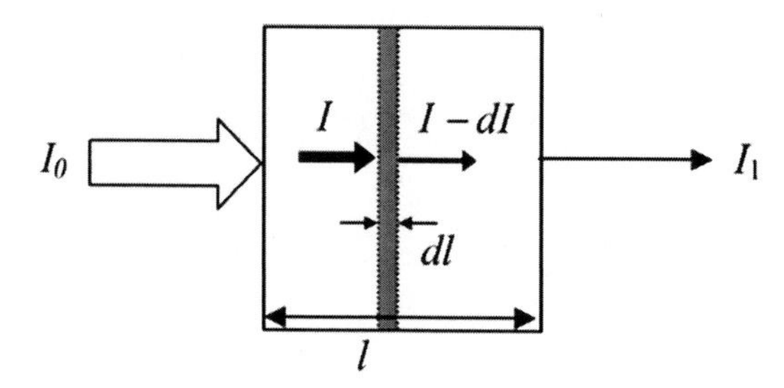

Fig. 1.1 Light transmission through a nonscattering medium

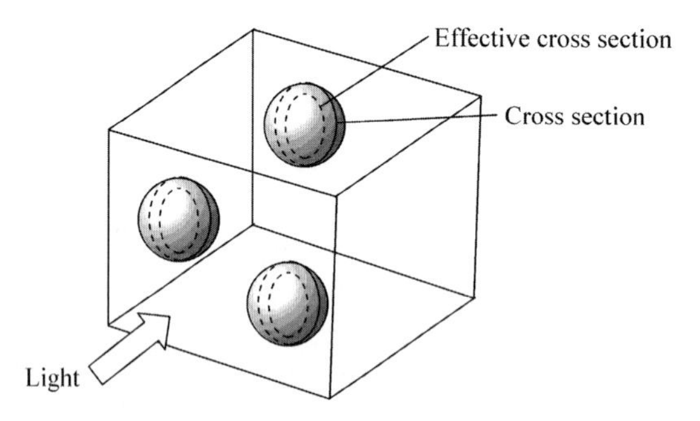

Fig. 1.2 Effective cross-section related to the scattering coefficient

When OD and ε are expressed as base 10 logarithms, the following expression is obtained from Eqs. 1.2, 1.3, and 1.4:

$$\mathrm{OD} = c\varepsilon l = \log_{10}(\mathrm{e})\ \mu_{\mathrm{a}} l, \tag{1.5}$$

where $\log_{10}(\mathrm{e})$ is 0.43429.

Scattering of light in biological tissue is caused by refractive index mismatches at boundaries, such as cell membranes and organelles. The area that contributes to scattering is called the effective cross-section, as depicted in Fig. 1.2. The scattering coefficient μ_{s} (mm^{-1}) is expressed as the cross-sectional area (mm^2) per unit volume of the medium (mm^3). When the scattered photon does not return to the incident axis, μ_{s} can be defined as

$$I = I_0 \exp(-\mu_{\mathrm{s}} l). \tag{1.6}$$

The scattering path length, defined as $1/\mu_{\mathrm{s}}$, is the expected value of distance that a photon travels between scattering events and defines the distance that reduces incident light I_0 to I_0/e due to scattering. The total attenuation coefficient, μ_{t} is defined as the sum of the absorption and scattering coefficients, and $1/\mu_{\mathrm{t}}$ is called the mean free path. When a photon is incident along the direction **i**, the angular probability of the photon being scattered in the direction **s** is given by the phase function $f(\mathbf{i},\mathbf{s})$. The phase function is conventionally expressed as a function of the cosine of the scattering angle. The anisotropy can be represented as the mean cosine of the scattering angle, and the anisotropy factor g is defined as

$$g = \int_{-1}^{1} \cos\theta\, f(\cos\theta) d\cos\theta. \tag{1.7}$$

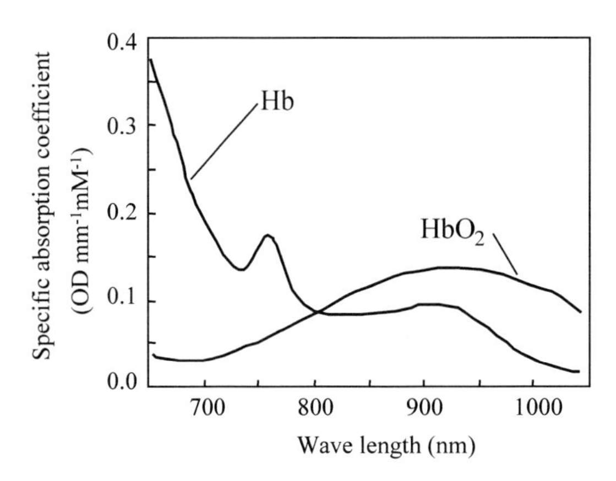

Fig. 1.3 Absorption spectra of oxyhemoglobin and deoxyhemoglobin

When $g = 0$ scattering is isotropic. Moreover, when $g = 1$ the incident light travels in a straight line. In contrast, when $g = -1$ complete backward scattering is observed. Biological tissues arestrongly forward-scattering media ($0.69 < g < 0.99$). The reduced scattering coefficient (μ'_s) is defined using the anisotropy factor as follows:

$$\mu'_s = \mu_s(1 - g). \tag{1.8}$$

The reduced scattering coefficient can be interpreted as representing the equivalent isotropic scattering coefficient and is used for the diffusion theory or Monte Carlo simulation when assuming isotropic scattering.

1.2 Optical Properties of Tissue

Hemoglobin is an iron-containing protein in red blood cells. One mole of deoxygenated hemoglobin (Hb) binds with four moles of oxygen to become oxygenated hemoglobin (HbO_2). The absorption spectra of oxygenated hemoglobin and deoxygenated hemoglobin [4] are shown in Fig. 1.3. The curves of the two hemoglobins intersect at about 800 nm, and the crossing point is called the isosbestic point. Since myoglobin and hemoglobin have similar absorption spectra, it is not easy to distinguish concentrations with spectroscopy. The separation of absorbers is also described in §1.4.

The absorption coefficient for water [5] is shown in Fig. 1.4. The absorption of water is small at wavelengths between about 200 and about 900 nm. Considering all components related to absorption in biological tissues, measurements at wavelengths between 680 and 950 nm are particularly suitable for spectroscopy.

Although many values for the optical properties of muscle and the overlying tissues (fat and skin) have been reported, there are significant differences in the results depending on the method of tissue preparation (fresh, saline-immersed, frozen, or thawed) and the theoretical analysis (diffusion theory, adding–doubling, Monte Carlo lookup tables). In Table 1.1 the optical properties of muscle, fat, dermis, and epidermis at wavelengths between 630 and 850 nm are given.

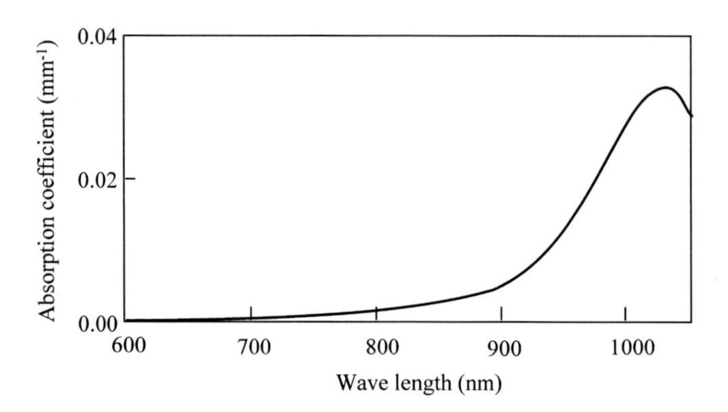

Fig. 1.4 Absorption spectrum of water

Table 1.1 Optical properties of muscle, fat, bone, dermis, and epidermis

Sample	λ (nm)	μ_a (mm^{-1})	μ_s' (mm^{-1})	References
Muscle				
Human forearm (in vivo)	800	0.015	1.0	Ferrari [6]
Human forearm (in vivo)	825	0.021–0.027	0.45–0.87	Zaccanti [7]
Human calf (in vivo)	825	0.018–0.028	0.51–0.85	Zaccanti [7]
Bovine muscle (in vitro)	633	0.096	0.53	Kienle [8]
Bovine muscle (in vitro)	751	0.037	0.34	Kienle [8]
Human calf (in vivo)	800	0.017 ± 0.005	0.80–1.1	Matcher [9]
Fat				
Human mamma (in vivo)	800	0.0017–0.0032	0.72–1.22	Mitic [10]
Human mamma (in vivo)	800	0.0023–0.0026	0.80–1.1	Suzuki [11]
Bovine fat (in vitro)	751	0.0021	1.0	Kienle [8]
Bone				
Pig skull (in vitro)	650	0.05	2.6	Firbank [12]
Pig skull (in vitro)	960	0.04	1.32	Firbank [12]
Human skull (in vivo)	849	0.022	0.91	Bevilacqua [13]
Dermis				
Pig dermis (in vitro)	790	0.018	1.4	Beek [14]
Pig dermis (in vitro)	850	0.033	0.9	Beek [14]
Epidermis				
Pig epidermis (in vitro)	790	0.24	1.9	Beek [14]
Pig epidermis (in vitro)	850	0.16	1.4	Beek [14]

1.3 Near-Infrared Spectroscopy

Spectroscopic measurement of in vivo tissue was first studied by Nicolai et al. [15] in 1932. They examined the optical characteristics of hemoglobin. The first practical ear oximeter for aviation use was developed by Millikan [16] ten years later. In 1949 Wood and Geraci [17] modified this instrument to obtain absolute values of oxygen saturation of arterial blood. The basic idea of this instrument was used to manufacture ear oximeters that were used in clinical settings until the 1970s. However, they did not have sufficient measurement stability for continuous monitoring of oxygen saturation because the calibration procedures were based on various extraneous assumptions. In 1974 Aoyagi et al. [18] presented a new idea called pulse oximetry, which utilizes the pulsation of arteries. This allowed for accurate measurement of oxygen saturation of arterial blood without the influence of

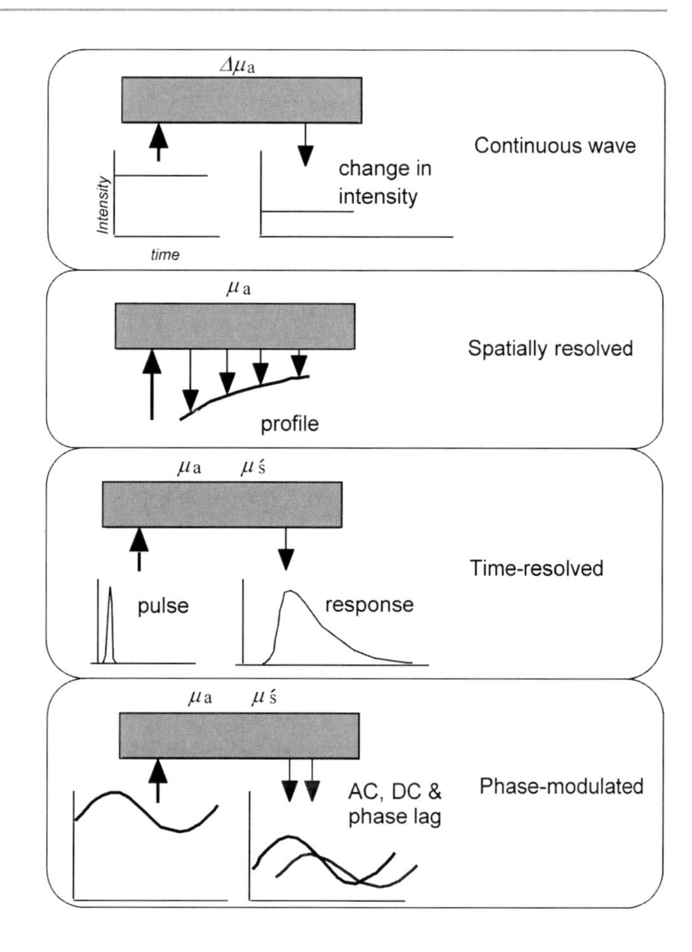

Fig. 1.5 Various techniques used in tissue oximetry employing NIRS

factors other than arterial blood. Pulse oximetry is now used in clinical medicine throughout the world. As mentioned above, oxygen saturation of arterial blood mainly reflects gas exchange occurring in the lungs and is an important factor for respiratory care. However, from the viewpoint of metabolism in tissues, measurement of blood oxygenation within the capillaries of each tissue is desirable.

In 1977, utilizing the relatively high penetration of near-infrared light, Jöbsis demonstrated that it was possible to measure attenuation spectra across the head of a cat, thereby obtaining information about tissue oxygenation [19]. In near-infrared spectroscopy (NIRS), tissue oxygenation is determined by analyzing the reflected or transmitted light intensity. However, NIRS using transmitted light is not suitable for clinical measurement, because it is difficult to detect transmitted light in adult human tissues; thus, reflection techniques are most commonly used nowadays. Several clinical studies on NIRS were conducted during the 1980s, including those by Ferrari et al. [20], Brazy et al. [21], Chance et al. [22], and Tamura et al. [23]. They detected changes in concentrations of oxy- and deoxyhemoglobin and in the cytochrome c oxidase redox state. They also demonstrated that NIRS is a useful, noninvasive technique for rapid detection of changes in tissue oxygenation.

Four major experimental techniques exist in the field of NIR spectroscopy, as shown in Fig. 1.5. The simplest one is continuous-wave spectroscopy (CWS) in which light of constant intensity is injected into tissue, and then the attenuated light signal is measured at a distance from the light source. The CWS technique has the limitation of obtaining only changes in optical density. More elaborate approaches are

Table 1.2 Advantages and disadvantages of CWS, SRS, TRS, and PMS

Parameters	CWS	SRS	TRS	PMS
$[HbO_2]$, [Hb], [tHb]	Changes	Absolute value	**Absolute value**	**Absolute value**
		(Assumed μ_s')*		
SO_2	No	**Yes**	**Yes**	**Yes**
Absorption coefficient	No	**Yes**	**Yes**	**Yes**
		(Assumed μ_s')[a]		
Scattering coefficient	No	No	**Yes**	**Yes**
Time-resolved profile	No	No	**Yes**	No
Mean path length	No	No	**Yes**	**Yes**
Sampling rate (Hz)	**≤100**	**≤100**	≤1	≤10
Portability	**Wearable**/portable	**Wearable**/portable	Portable	Portable
Instrument cost	Low/moderate	Low/moderate	High	Moderate
Initial stabilization	**Not required**	**Not required**	Required	**Not required**
Light source	LED/laser diode	LED/laser diode	Laser diode	Laser diode
Detector	Silicon photodiode	Silicon photodiode	Photomultiplier tube	Avalanche photodiode

Underlined character = advantage, Bold character = important advantage, LED = light emitting diode. [a] = the calculation is possible in principle, but whether the parameter is calculated or not depends on the instrumentation

spatially resolved spectroscopy (SRS), time-resolved spectroscopy (TRS), and phase-modulated spectroscopy (PMS). Table 1.2 shows the advantages and disadvantages of the four measurement methods. The principles of these techniques are described in the following section.

1.4 Continuous-Wave NIRS

In NIRS, scattered light is detected at a distance from the light source, and tissue oxygenation is determined from the change in absorption coefficients of a tissue using the basic equations of conventional oximetry. Oximetry is the colorimetric measurement of the degree of oxygen saturation. Assuming that changes in light absorption are mainly due to changes in blood oxygenation or volume, $[HbO_2]$ and [Hb] can be determined as follows. Change in the absorption coefficient of a tissue $\Delta\mu_a$ is expressed as

$$\Delta\mu_a = \varepsilon_\lambda^{HbO_2}\Delta[HbO_2] + \varepsilon_\lambda^{Hb}\Delta[Hb] \tag{1.9}$$

where $\varepsilon_\lambda^{HbO_2}$ and ε_λ^{Hb} are molar absorption coefficients of HbO_2 and Hb at wavelength λ, respectively. For example, ε_λ^{Hb} at a wavelength of 760 nm is 0.1674 OD mM^{-1} mm^{-1}, as reported by Matcher et al. [4]. In the above equations, ε_λ^{Hb} of 0.385 mM^{-1} mm^{-1} (= 0.1674 × ln10) is used because Matcher et al. defined OD as the logarithm to base 10 and μ_a is defined in base e. The two unknowns, $\Delta[HbO_2]$ and Δ[Hb], are obtained from measurements at two wavelengths. The NIRS instruments usually use a combination of wavelengths between 680 and 950 nm. These wavelengths are usually chosen to be around an isosbestic point (805 nm) of HbO_2 and Hb – for example, 770/830, 760/840, and 690/900 nm. When the difference of two wavelengths is large, changes in intensity due to wavelength are easily obtained, but the change in optical path length would not be ignored. The following equations are solved on assumption that the path lengths of each wavelength are same:

$$\Delta[\mathrm{HbO_2}] = \frac{1}{k}\left(\varepsilon_2^{\mathrm{Hb}}\Delta\mu_{\mathrm{a1}} - \varepsilon_1^{\mathrm{Hb}}\Delta\mu_{\mathrm{a2}}\right), \quad (1.10)$$

$$\Delta[\mathrm{Hb}] = \frac{-1}{k}\left(\varepsilon_2^{\mathrm{HbO_2}}\Delta\mu_{\mathrm{a1}} - \varepsilon_1^{\mathrm{HbO_2}}\Delta\mu_{\mathrm{a2}}\right), \quad (1.11)$$

$$k = \varepsilon_1^{\mathrm{HbO_2}}\varepsilon_2^{\mathrm{Hb}} - \varepsilon_1^{\mathrm{Hb}}\varepsilon_2^{\mathrm{HbO_2}}. \quad (1.12)$$

The change in absorption, $\Delta\mu_a$, can be determined by various NIRS techniques, such as CWS, SRS, PMS, and TRS. The CWS method, which is the simplest one, only enables determination of change in absorption. In contrast, the optical properties in absolute values can be obtained by using SRS, PMS, or TRS. However, with any technique it is difficult to quantify the concentration other than that of hemoglobin, such as myoglobinin muscle tissue, cytochrome oxidase, carboxy-hemoglobin, and methemoglobin. Schenkman et al. [24, 25] reported a method for quantification of myoglobin and hemoglobin using the difference in peak position around 760 nm at near-infrared wavelengths. A composite myoglobin–hemoglobin peak would be slightly shifted by the absorption of hemoglobin derivatives and path length for each wavelength. Alternatively, some researchers have used NMR to observe the distinct Mb and Hb signals and then apply the results to determine Mb and Hb contributions in the NIRS spectra [26, 27].

Further studies with accurate quantification of small-quantity absorbers, mentioned above, scattering coefficient (or path length) for each wavelength, and combination of spectroscopic techniques could potentially form a basis for developing key technologies to measure the myoglobin and hemoglobin contribution to the NIRS signal.

In CW-NIRS, change in the optical density – defined by $\Delta\mathrm{OD} = \ln(R_0/R)$, where R_0 and R are intensities of backscattered light at a reference state (usually taken at the start of measurement) and during measurement, respectively – is measured. Assuming that the scattering coefficient does not change during measurement, we can determine $\Delta\mu_a$ using the modified Beer–Lambert law: $\Delta\mathrm{OD} = \Delta\mu_a d$, where $d = \partial\mathrm{OD}/\partial\mu_a$, which is defined as the differential path length and is equal to the mean optical path length.

CW-NIRS systems measure only ΔOD, and at least two different wavelengths are usually employed to obtain spectral information. Relative changes of HbO_2 and Hb are continuously monitored utilizing Eqs. 1.10 and 1.11. The CW method is advantageous because it is highly sensitive, enabling a data sampling rate of less than a second, economical, and can be miniaturized to the extent of a multipoint monitor even for imaging. The assembly of a multiwavelength, multisource, multidetector imager for brain function and the circuit diagram are depicted in Fig. 1.6 [28].

Silicon photodiodes and multiwavelength light-emitting diodes (LEDs) are used as detectors and the near-infrared light source, respectively. They are held by elastic bands at a source–detector distance of 3 cm. The penetration depth of light and the spatial resolution are about 1.5 and 2 cm, respectively. Because digital gain control is used, the system can readily be controlled over a 20-dB range to equalize 16-channel signals. The output is then connected to a multiplexer (MUX) switch, which is synchronized with the flashing of the LED, so that one wavelength is sampled by separate integrating capacitors, which gives an RC charging time constant. The stored signal in this capacitor can be updated step by step by using the MUX switch. The storage capacitor is then oversampled by an analog-to-digital converter (ADC) at 250 samples/s to avoid aliasing. The temporal resolution of oxygenation measurement is $\geq$0.3 s.

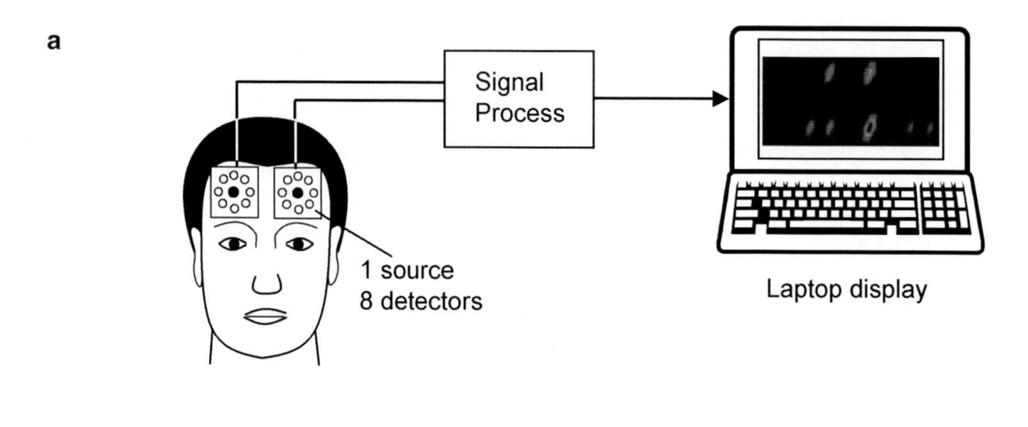

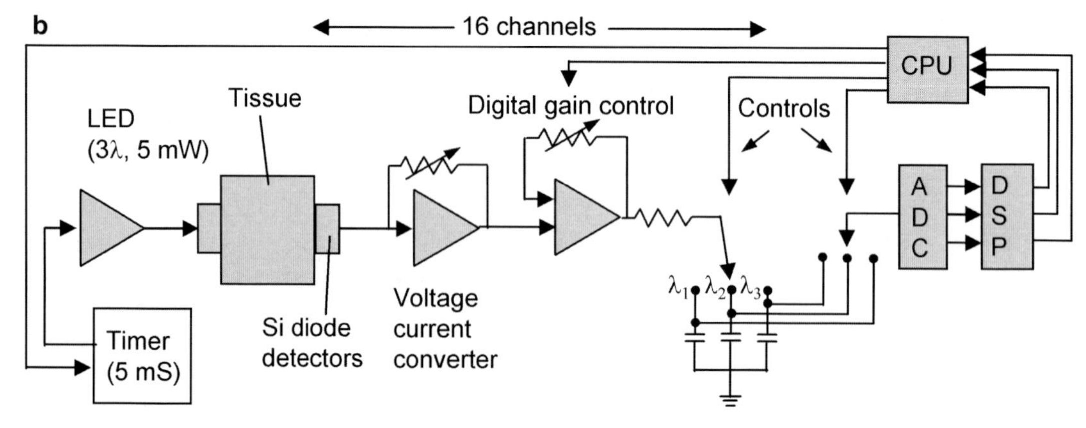

Fig. 1.6 The 16-channel "CW imager" is illustrated in (**a**) giving circuit constants and component values for one channel in (**b**)

Brain activation was monitored during the anagram test, which required the subject to identify a five-letter word from some items. The imager pad is centered on the nose bridge and symmetrically attached to the forehead, eyebrow to hairline, and temple to temple. The covering area is corresponding to Brodmann's areas 9 and 10, which are the part of the frontal cortex in the human brain. The subject's signals are illustrated in Fig. 1.7. The data show postsolution hyperemia and brief deoxygenation prior to problem solving and prolonged hyperoxygenation thereafter.

Although the CW method gives only relative values, it is sufficient for many cases, such as studies of the functional activity of the brain [29–32] or interventional studies for testing reactions on drugs or changes in treatment.

1.5 Spatially Resolved NIRS

Patterson et al. [33] proposed that the effective attenuation coefficient μ_{eff} of tissues can be obtained by measuring the spatial profile of the intensity of backscattered light as a function of the distance from the light source using a large source–detector separation. They showed that the intensity of reflected light R can be expressed as follows:

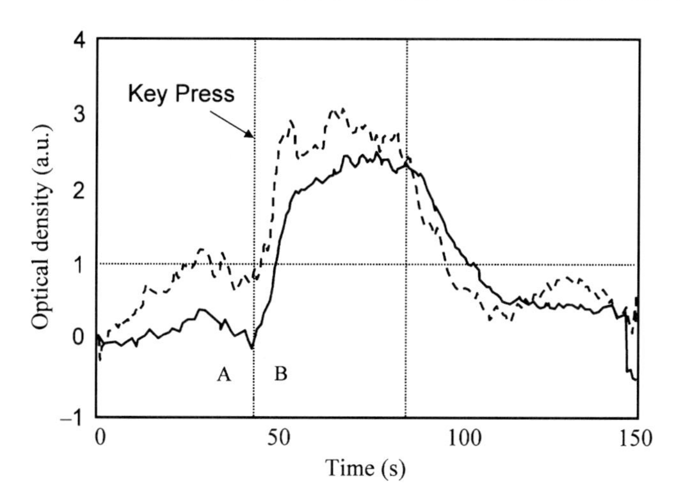

Fig. 1.7 Raw waveform data obtained during an anagram test. *Dashed line*: blood volume; *solid line*: oxygenation. The decision that the anagram has been solved is indicated by the *dashed vertical line* labeled "Key press"

$$R \propto \frac{\exp(-\mu_{\text{eff}})}{\rho^2}, \tag{1.13}$$

where $\mu_{\text{eff}} = \sqrt{3\mu_{\text{a}}(\mu_{\text{a}} + \mu'_{\text{s}})}$.

In the case of CW spectroscopy, the local reflectance $I(\rho)$ at position ρ is expressed as an integral of $R(\rho, t)$ over time, and OD is defined as the negative of the logarithm of $I(\rho)$.

Thus, OD is expressed by

$$\begin{aligned} OD &= -\ln I(\rho) \\ &= -\ln(z_0/2\pi) + \frac{3}{2}\ln\left(\rho^2 + z_0{}^2\right) - \ln\left(1 + \mu_{\text{eff}}\left(\rho^2 + z_0{}^2\right)^{1/2}\right) + \mu_{\text{eff}}\left(\rho^2 + z_0{}^2\right)^{1/2}. \end{aligned} \tag{1.14}$$

In the measurement of human tissues, if $\rho^2 \gg z_0{}^2$ is assumed, $\mu_{\text{eff}}\,\rho \gg 1$, the following equation is derived.

$$\text{OD} = -\ln(z_0/2\pi) + 3\ln(\rho) - \ln(\mu_{\text{eff}}\rho) + \mu_{\text{eff}}\rho. \tag{1.15}$$

Differentiating OD with respect to ρ and assuming that $\mu_{\text{a}} \ll \mu_{\text{s}}'$ yield the following relation [34, 35]:

$$\frac{\partial \text{OD}}{\partial \rho} = -\frac{\partial}{\partial \rho}\ln I(\rho) = \frac{2}{\rho} + \sqrt{3\mu_a\mu_s'},$$

$$\therefore \mu_a\mu_s' = \frac{1}{3}\left(\frac{\partial \text{OD}}{\partial \rho} - \frac{2}{\rho}\right)^2. \tag{1.16}$$

where $\partial\text{OD}/\partial\rho$ is the local gradient of attenuation with respect to the source–detector separation.

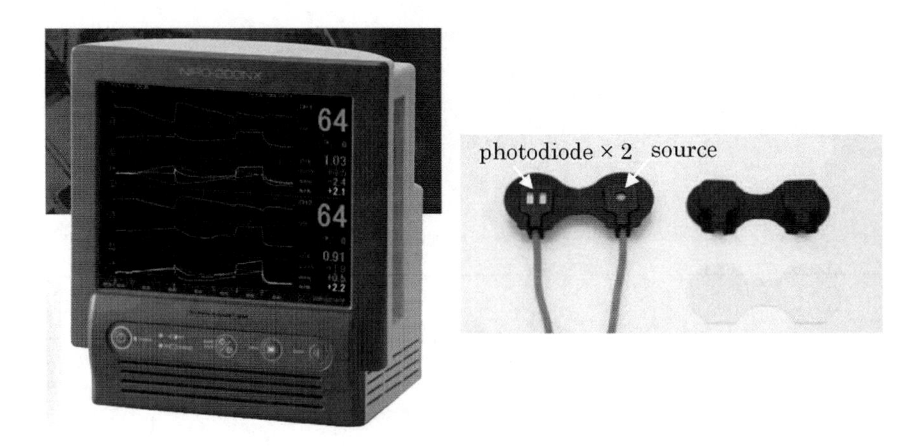

Fig. 1.8 Spatially resolved spectroscopy system (NIRO-200NX) with an optical probe consisting of photodiodes and a light source

In the first-order approximation, μ'_s can be assumed to be constant within a narrow wavelength region of NIR light. Then the relative concentration changes of HbO_2 and Hb are derived from the following equation:

$$\begin{bmatrix} \left(\frac{\partial \mathrm{OD}(\lambda_1)}{\partial \rho}\right)^2 - \frac{2}{\rho} \\ \left(\frac{\partial \mathrm{OD}(\lambda_2)}{\partial \rho}\right)^2 - \frac{2}{\rho} \\ \left(\frac{\partial \mathrm{OD}(\lambda_3)}{\partial \rho}\right)^2 - \frac{2}{\rho} \end{bmatrix} = 3\mu'_s \begin{bmatrix} \varepsilon_{\mathrm{HbO_2}}(\lambda_1) & \varepsilon_{\mathrm{Hb}}(\lambda_1) \\ \varepsilon_{\mathrm{HbO_2}}(\lambda_2) & \varepsilon_{\mathrm{Hb}}(\lambda_2) \\ \varepsilon_{\mathrm{HbO_2}}(\lambda_3) & \varepsilon_{\mathrm{Hb}}(\lambda_3) \end{bmatrix} \begin{bmatrix} C_{\mathrm{HbO_2}} \\ C_{\mathrm{Hb}} \end{bmatrix}. \tag{1.17}$$

Moreover, the tissue oxygenation index (TOI) is calculated using the following equation:

$$\mathrm{TOI} = \frac{C_{\mathrm{HbO_2}}}{C_{\mathrm{HbO_2}} + C_{\mathrm{Hb}}} \times 100\ (\%). \tag{1.18}$$

A commercially available instrument using SRS for the measurement of hemoglobin saturation has been developed by Hamamatu Photonics, as shown in Fig. 1.8 [35].

A NIRO 300 was incorporated into an established multimodal monitoring system, enabling recording of cerebral hemodynamic changes during carotid endarterectomy (CEA) [36]. Brief periods of cerebral ischemia often occur during cross-clamping of the internal carotid artery (ICA) during surgery. Multimodal monitoring consists of frontal cutaneous laser-Doppler flowmetry (LDF) and transcranial Doppler mean flow velocity (FV) measurements of the ipsilateral middle cerebral artery. Typical data obtained during CEA are shown in Fig. 1.9. Sequential clamping was performed on the external carotid artery (ECA) before ICA clamping. The measurements obtained by LDF can be seen to fall only when the ECA clamp is applied. In this case the drop in FV is seen to be specific to ICA clamping, similar to the drop in the TOI. On insertion of an ICA vascular shunt, FV, and TOI were restored to values approaching baseline levels.

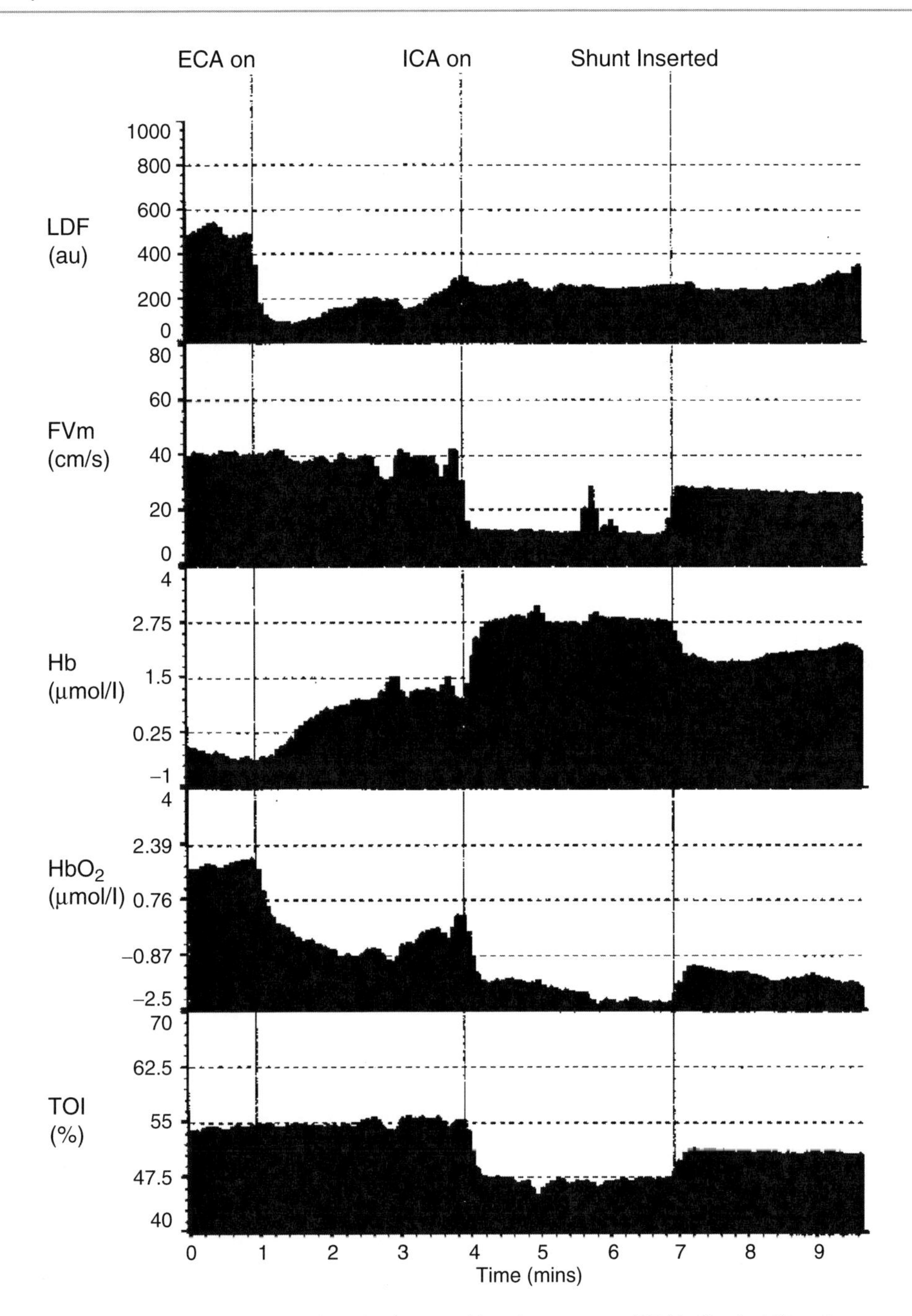

Fig. 1.9 Data obtained from a patient during elective carotid endarterectomy (CEA). *Vertical lines* demonstrate time of application of vascular clamps

1.6 Time-Resolved NIRS

In TRS temporal changes in the reflected light intensity are measured after irradiation of a picosecond pulse, thereby giving a distribution of the total path length of a photon traveling in the scattering medium [6, 37–39]. This technique can be used to determine the absorption coefficient and the

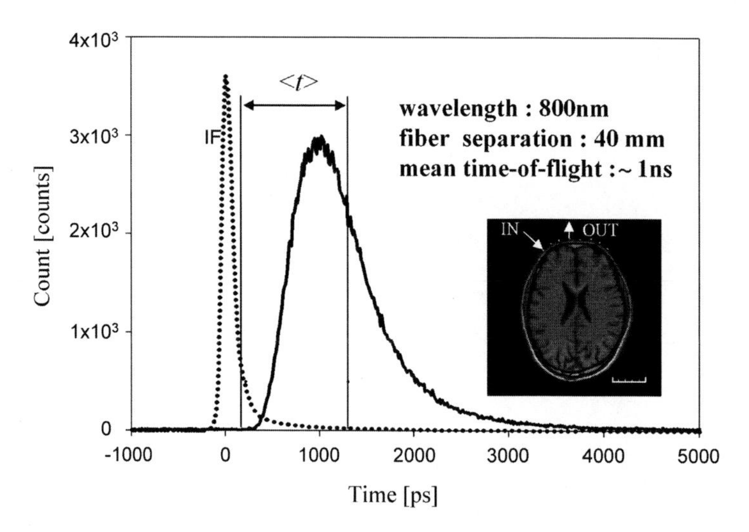

Fig. 1.10 Time-resolved waveform of the incident short pulse and reflectance at a 40-mm separation

reduced scattering coefficient of tissues. A method for determining the absorption and scattering coefficients is based on a curve fitting between measured data and a theoretical curve obtained by diffusion theory. When a semiinfinite medium is assumed and the zero-boundary condition is applied, the reflectance R at a source–detector separation ρ and time t is given [40] by

$$R(\rho,t) = (4\pi cD)^{-3/2}(\mu_a + \mu_s{}')^{-1}t^{-5/2}\exp\left(-\frac{\rho^2 + (\mu_a + \mu_s{}')^{-2}}{4cD}\frac{1}{t}\right)\exp(-\mu_a ct). \tag{1.19}$$

Taking the natural logarithm on both sides of Eq. 1.19 and assuming that $\rho \gg (\mu_\text{a} + \mu'_\text{s})^{-1}$,we obtain the following equation:

$$\ln[R(\rho,t)] = \kappa - \frac{5}{2}\ln(t) - \left(ct + \frac{3\rho^2}{4ct}\right)\mu_\text{a} - \frac{3\rho^2}{4ct}\mu_\text{s}{}', \tag{1.20}$$

where $\kappa = -\ln(4\pi cD)3/2 - \ln(\mu_\text{a} + \mu_\text{s}{}')$. Simple mean least-squares-fitting algorithms can be used to determine μ_a and $\mu_\text{s}{}'$ from experimental data. A typical waveform of time-resolved measurement is shown in Fig. 1.10.

A TRS system (TRS-20) uses the time-correlated single-photon counting (TCPC) method to measure the temporal profile of the detected photons (see Fig. 1.11). The system [41] consists of a three-wavelength (759, 797, and 833 nm) light pulse source (PLP: Picosecond Light Pulser, Hamamatsu Photonics KK, Hamamatsu, Japan), which generates light pulses with a peak power of about 60 mW, pulse width of 100 ps, pulse rate of 5 MHz, and an average power of 30 μW for one wavelength. For the detection, a photomultiplier tube (PMT, H7422-50MOD, Hamamatsu Photonics KK) was used in photon-counting mode. The timing signals were received and accumulated by a TRS circuit that consists of a constant-fraction discriminator, a time-to-amplitude converter, an ADC, and histogram memory. The PLP emits three-wavelength light pulses in turn, and the light pulses are guided into one illuminating optical fiber by a fiber coupler (CH20G-D3-CF, Mitsubishi Gas Chemical Company Inc., Japan). A single optical fiber (GC200/250L, FUJIKURA Ltd., Japan) with a numerical aperture (NA) of 0.21 and a core diameter of 200 μm was used for illumination.

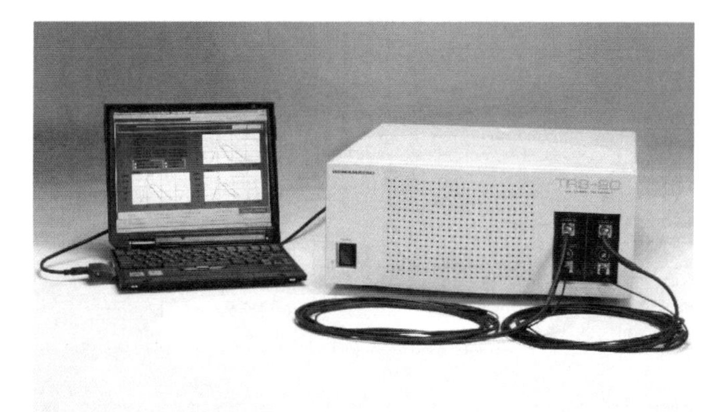

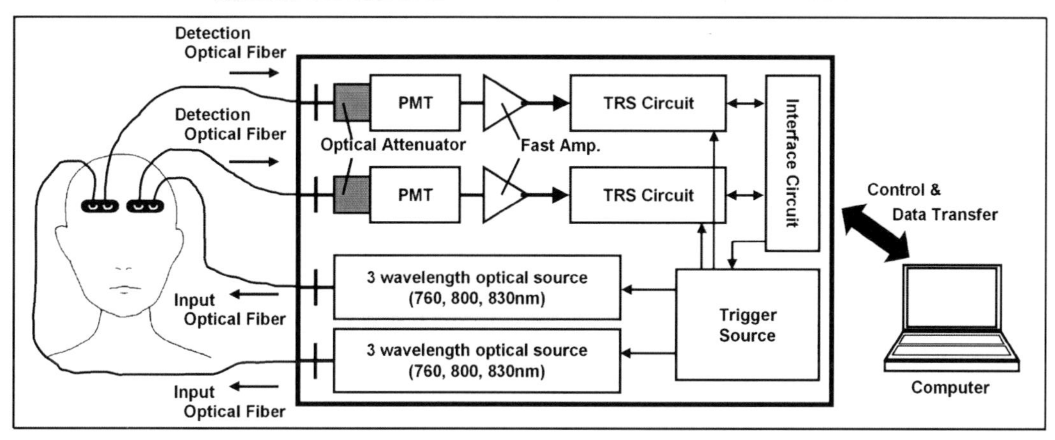

Fig. 1.11 Schematic diagram of a time-resolved spectroscopy system

An optical bundle fiber (LB21E, Moritex Corp., Japan) with an NA of 0.21 and a bundle diameter of 3 mm was used to collect diffuse light from the tissues. TRS-20 has two sets of PLP and TCPC detectors, enabling the independent measurement of two portions.

TRS allows for determination of relative light intensity, mean optical path length, transport scattering coefficient (μ'_s), and μ_a. The intensity can be obtained by integrating the temporal profiles, and the modified Beer–Lambert law uses this information to calculate absorbance changes. The mean optical path lengths were calculated from the center of gravity of the temporal profile [42]. The calculations of the intensity, absorbance change, and mean path length are model independent. Applying the diffusion equation for semiinfinite homogeneous media with zero-boundary conditions in reflectance mode into all observed temporal profiles, we obtained the values of μ'_s and μ_a using the nonlinear least-squares method [43].

If it is assumed that absorption in the 700–900 nm range arises from absorption of HbO_2, Hb, and water, $\mu_{a\lambda}$ of the measured wavelengths: λ (759, 797, and 833 nm) is expressed as shown in simultaneous Eq. 1.21 [44, 45]:

$$\begin{aligned}
\mu_{a759nm} &= \varepsilon_{HbO_2 759nm} C_{HbO_2} + \varepsilon_{Hb759nm} C_{Hb} + \mu_{aH_2O759nm} \\
\mu_{a797nm} &= \varepsilon_{HbO_2 797nm} C_{HbO_2} + \varepsilon_{Hb797nm} C_{Hb} + \mu_{aH_2O797nm} \\
\mu_{a833nm} &= \varepsilon_{HbO_2 833nm} C_{HbO_2} + \varepsilon_{Hb833nm} C_{Hb} + \mu_{aH_2O833nm},
\end{aligned} \tag{1.21}$$

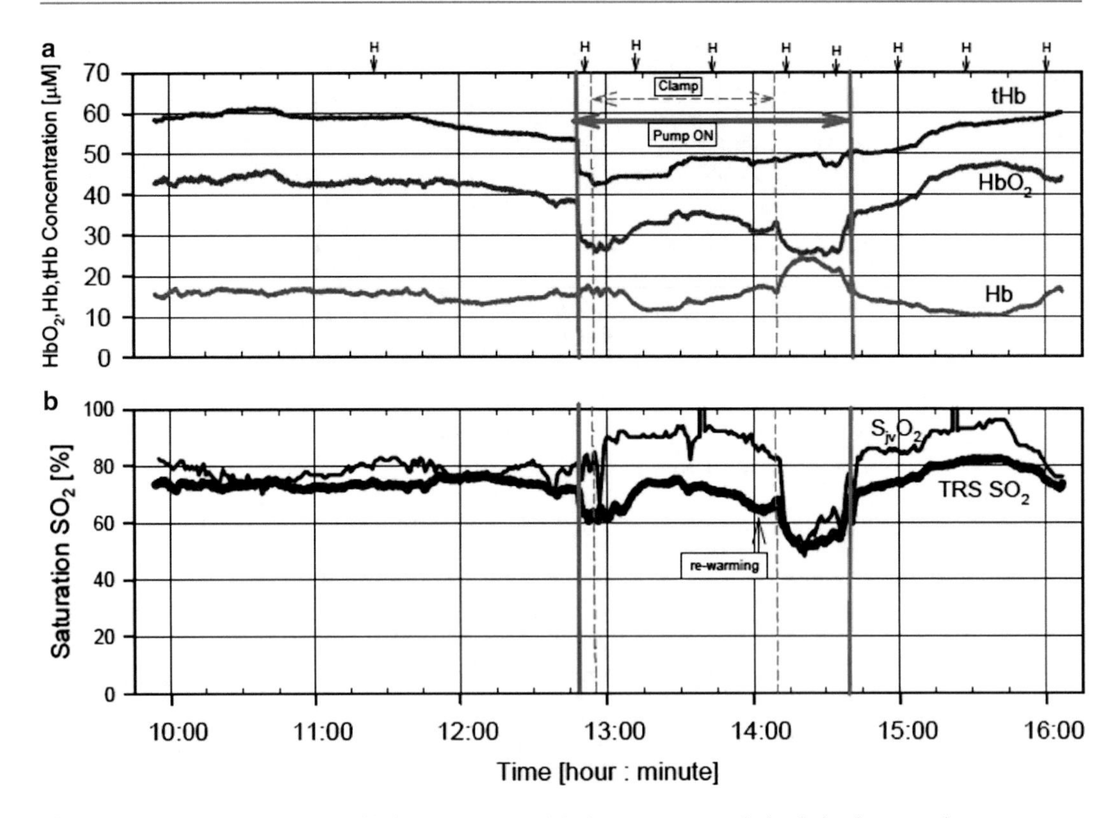

Fig. 1.12 Fluctuations of SO_2 and $SjvO_2$ were separated during extracorporeal circulation in one patient

where $\varepsilon_{m\lambda}$ is the molar absorption coefficient of substance m at wavelength λ, and C_m is the concentration of substance m. After subtracting water absorption from μ_a at each wavelength, assuming that the volume fraction of the water content was constant, $[HbO_2]$ and [Hb] were determined using the least-squares-fitting method.

The total concentrations of Hb (tHb) and tissue oxygen saturation (SO_2) were calculated as follows:

$$\mathrm{tHb} = \mathrm{Hb}_{\mathrm{O}_2} + \mathrm{Hb}, \tag{1.22}$$

$$\mathrm{SO}_2 = \frac{[\mathrm{HbO}_2]}{[\mathrm{tHb}]} \times 100. \tag{1.23}$$

The hemodynamics of the brain were monitored by TRS during a coronary artery bypass grafting surgery using an artificial heart-and-lung machine [45]. Figure 1.12 shows the time course of changes in $[HbO_2]$, [Hb], and [tHB] in (A) and SO_2 and the internal jugular vein oxygen saturation ($SjvO_2$) in (B). When extracorporeal circulation was started by the pump, HbO_2 and tHb decreased rapidly. At the end of extracorporeal circulation, those values returned to the initial levels. SO_2 estimated by TRS was found to be nearly the same as $SjvO_2$ before and after extracorporeal circulation, but during circulation they behaved differently in this case.

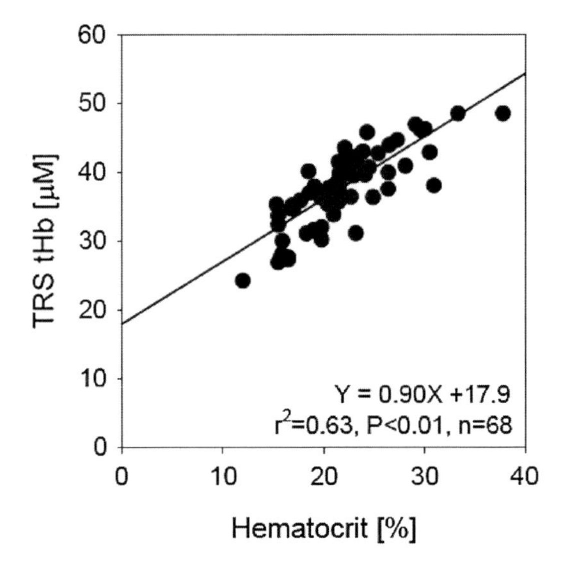

Fig. 1.13 Relationship between tHb by TRS-10 and hematocrit (Hct) in nine patients

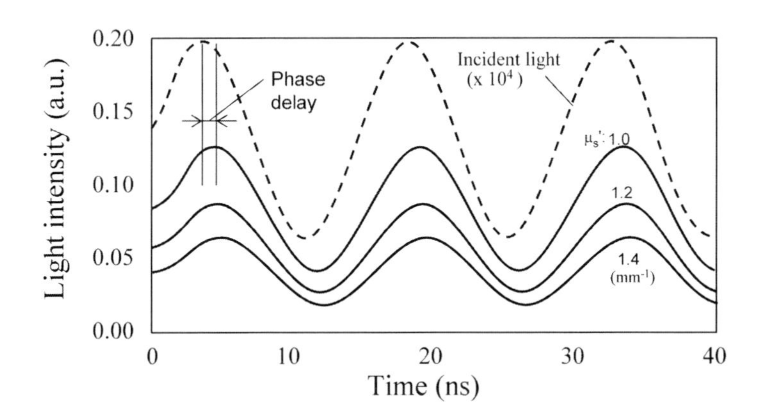

Fig. 1.14 Phase shift between the incident light (*dashed line*) and scattered light through tissues (*solid lines*) at 70 MHz of modulation frequency

Figure 1.13 shows the correlation between the hematocrit (Hct) values of arterial blood and tHb in 9 patients. The correlation among patients was high ($r^2 = 0.63$), showing that tHb measured by TRS has good linearity with Hct.

1.7 Phase-Modulated NIRS

Phase-modulated (frequency domain) measurements were first reported by Chance in 1949 [46]. Pulse code or phase modulation gives mean time delay between source and detector. The time delay is related to light scattering and absorption, including biological signals. Figure 1.14 shows an example of the phase shift obtained by a computer simulation. Different types of equipment were developed

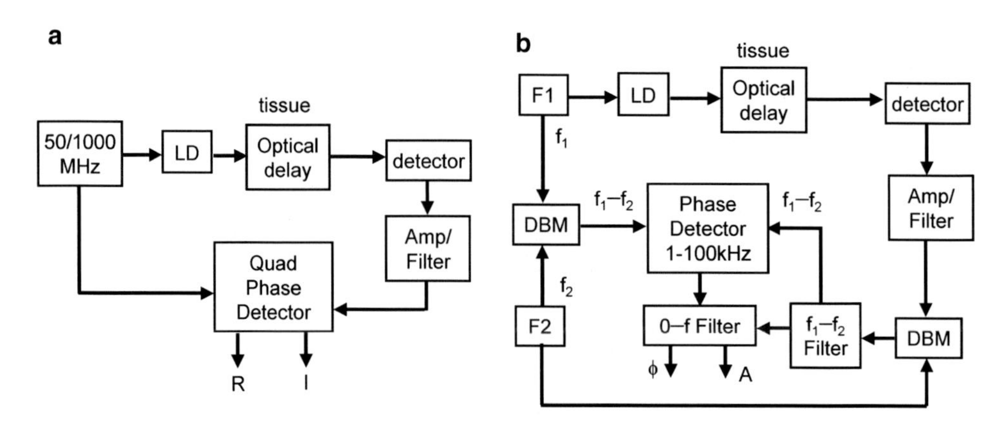

Fig. 1.15 Schematic diagram of phase-modulated homodyne (**a**) and heterodyne (**b**) NIRS systems

and classified by the multiplicity of wavelengths and the type of phase detection. Homodyne systems [47], heterodyne systems [48, 49], and network analyzers [50] were used for measuring the phase (see Fig. 1.15). The performance of both homodyne and heterodyne detection systems was examined, and it was suggested that a homodyne system has advantages in terms of simplicity of construction and execution, while a heterodyne system has high precision and the possibility of low-frequency phase detection [51].

The supporting theoretical background is the diffusion equation. An analytical solution was obtained on the basis of an assumption that the modulation frequency ($\omega/2\pi$) is much smaller than the typical frequency of scattering processes (i.e., $c\mu'_s$, where c is the speed of light in a medium). The condition for this assumption is satisfied by most biological tissues in an NIR spectral region for modulation frequencies up to 1 GHz. Fishkin and Gratton obtained the following expressions [52]:

$$\ln(\rho U_{\mathrm{dc}}) = -\rho\sqrt{\frac{\mu_{\mathrm{a}}}{D}} + \ln\left(\frac{S}{4\pi D}\right), \tag{1.24}$$

$$\ln(\rho U_{\mathrm{ac}}) = -\rho A\cos\left(\frac{\theta}{2}\right) + \ln\left(\frac{kS}{4\pi D}\right), \tag{1.25}$$

$$\phi = \rho A\sin\frac{\theta}{2}, \tag{1.26}$$

$$\theta = \tan^{-1}\left(\frac{\omega}{\mu_{\mathrm{a}}c}\right), \quad A = \left(\frac{\mu_{\mathrm{a}}^2c^2 + \omega^2}{c^2D^2}\right)^{1/4}, \tag{1.27}$$

where ϕ is the phase shift of a detected signal relative to an excited signal, U_{dc} is the direct current (dc) component of the photon density, U_{ac} is the amplitude of the alternating current (ac) component of the photon density, S is the source strength (in photons per second), and k is defined as the ratio of the ac to the dc components of the intensity. Fantini et al. [53] developed practical instrumentation based on diffusion theory with multidistance detection. As mentioned above, various phase-modulated NIRS systems have been developed. A typical commercial phase-modulated NIRS system is depicted in Fig. 1.16. OxiplexTS (ISS Inc., Champaign, IL) is a tissue oximeter that includes light sources (690 and 830 nm), with a multidistance emitter array, and one detection channel using PMT.

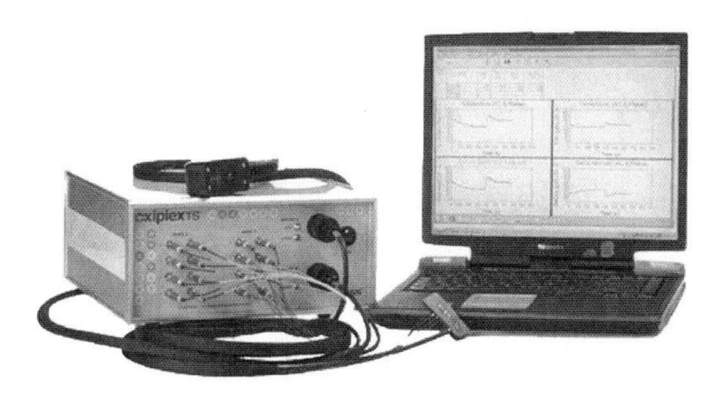

Fig. 1.16 Phase-modulated NIRS system, ISS Oxiplex (ISS Inc.)

The detection type is heterodyne with an offset frequency of several kHz at a 110-MHz modulation frequency. Various optical probes can be coupled to the system for specific medical research applications.

Problem

1.1. How can the weak photocurrent of an Si photodiode be converted to a voltage signal on continuous-wave NIRS or spatially resolved NIRS?

Further Reading

Demrow B (1971) Op amps as electrometers or — the world of fA. Anal Dial 5(2): 48–49

Frenzel LE (2007) Accurately measure nanoampere and picoampere currents. Electron Design Strat News, Feb 15

Hutchings MJ, Blake-Coleman BC (1994) A transimpedance converter for low-frequency, high-impedance measurements. Meas Sci Technol 5(3):310–313

Rako P (2007) Measuring nanoamperes. Electron Design Strat News, Apr 26

References

1. Bouguer P (1729) Essai d'optique sur la gradation de la lumière. Claude Jombert, Paris
2. Lambert JH (1760) Lambert's photometrie: photometria, sive de mensura et gradibus luminis, colorum et umbrae. Wilhelm Engelmann, Berlin
3. Beer A (1852) Bestimmung der absorption des rothen Lichts in farbigen Flüssigkeiten. Annu Rev Phys Chem 86:78–88
4. Matcher SJ, Elwell CE, Cooper CE, Cope M, Delpy DT (1995) Performance comparison of several published tissue near-infrared spectroscopy algorithms. Anal Biochem 227:54–68
5. Hale GM, Querry MR (1973) Optical constants of water in the 200-nm to 200-mm wavelength region. Appl Opt 12:555–563
6. Ferrari M, Wei Q, Carraresi L, De Blasi RA, Zaccanti G (1992) Time-resolved spectroscopy of the human forearm. J Photochem Photobiol B: Biol 16:141–153
7. Zaccanti G, Taddeucci A, Barilli M, Bruscaglioni P, Martelli F (1995) Optical properties of biological tissues. Proc SPIE 2389:513–521
8. Kienle A, Lilge L, Patterson MS, Hibst R, Steiner R, Wilson BC (1996) Spatially resolved absolute absorption coefficients of biological tissue. Appl Opt 35:2304–2314
9. Matcher SJ, Cope M, Delpy DT (1997) In vivo measurements of the wavelength dependence of tissue-scattering coefficients between 760 and 900 nm measured with time-resolved spectroscopy. Appl Opt 36:386–396

10. Mitic G, Közer J, Otto J, Plies E, Sökner G, Zinth W (1994) Time-gated transillumination of biological tissues and tissue like phantoms. Appl Opt 33:6699–6710
11. Suzuki K, Yamashita Y, Ohta K, Chance B (1994) Quantitative measurement of optical parameters in the breast using time-resolved spectroscopy phantom and preliminary in vivo results. Invest Radiol 29:410–414
12. Firbank M, Hiraoka M, Essenpreis M, Delpy DT (1993) Measurement of the optical properties of the skull in the wavelength range 650–950 nm. Phys Med Biol 38:503–510
13. Bevilacqua F, Piguet D, Marquet P, Gross JD, Tromberg BJ, Depeursinge C (1999) In vivo local determination of tissue optical properties: applications to human brain. Appl Opt 38:4939–4950
14. Beek JF, van Staveren HJ, Posthumus P, Sterenborg HJ, van Gemert MJ (1993) The influence of respiration on optical properties of piglet lung at 632.8 nm. Med Opt Tomogr 32:193–210
15. Nicolai L (1932) Über sichtbarmachung, verlauf und chemische kinetic der, oxyhemoglobinreduktion im lebendum gewebe, besonders in der menschlichen haut. Arch Gesch Physiol 229:372–384
16. Millikan GA (1942) The oximeter, an instrument for measuring continuously oxygen saturation of arterial blood in man. Rev Sci Instrum 13:434–444
17. Wood EH, Geraci JE (1949) Photoelectric determination of arterial oxygen saturation in man. J Lab Clin Invest 34:387–401
18. Aoyagi T, Kishi M, Yamaguchi K, Watanabe S (1974) Improvement of earpiece oximeter. Proc 13th Conf Jpn Soc Med Electron Biol Eng 12:90–91
19. Jöbsis FF (1977) Noninvasive, infrared monitoring of cerebral and myocardial oxygen sufficiency and circulatory parameters. Science 198:1264–1267
20. Ferrari M, Giannini I, Sideri G, Zanette E (1985) Continuous noninvasive monitoring of human brain by near infrared spectroscopy. Adv Exp Med Biol 191:873–882
21. Brazy JE, Lewis DV, Mitnick MH, Jöbsis FF (1985) Noninvasive monitoring of cerebral oxygenation in preterm infants. Pediatrics 75:217–225
22. Chance B, Nioka S, Kent J, McCully K, Fountai M, Greenfeld R, Holtom G (1988) Time-resolved spectroscopy of hemoglobin and myoglobin in resting and ischemic muscle. Anal Biochem 174:698–707
23. Tamura M, Hazeki O, Nioka S, Chance B, Smith DS (1988) The simultaneous measurements of tissue oxygen concentration and energy state by near-infrared and nuclear magnetic resonance spectroscopy. Adv Exp Med Biol 222:359–363
24. Schenkman KA, Marble DA, Feiglf EO, Burns DH (1999) Near-infrared spectroscopic measurement of myoglobin oxygen saturation in the presence of hemoglobin using partial least-squares analysis. Appl Spectrosc 53:325–331
25. Marcinek DJ, Amara CE, Matz K, Conley KE, Schenkman KA (2007) Wavelength shift analysis: a simple method to determine the contribution of hemoglobin and myoglobin to in vivo optical spectra. Appl Spectrosc 61:665–669
26. Tran TK, Sailasuta N, Kreutzer U, Hurd R, Chung Y, Mole P, Kuno S, Jue T (1999) Comparative analysis of NMR and NIRS measurements of intracellular PO_2 in human skeletal muscle. Am J Physiol 276:R1682–R1690
27. Xie H, Kreutzer U, Jue T (2009) Oximetry with the NMR signals of hemoglobin Val E11 and Tyr C7. Eur J Appl Physiol 107:325–333
28. Chance B, Nioka S, Zhao Z (2007) A wearable brain imager. IEEE Eng Med Biol 26:30–37
29. Hoshi Y, Tamura M (1993) Detection of dynamic changes in cerebral oxygenation coupled to neuronal function during mental work in man. Neurosci Lett 150:5–8
30. Chance B, Zhuang Z, UnAh C, Alter C, Lipton L (1993) Cognition-activated low-frequency modulation of light absorption in human brain. Proc Natl Acad Sci USA 90(8):3770–3774
31. Kato T, Kamei A, Takashima S, Ozaki T (1993) Human visual cortical function during photic stimulation monitoring by means of near-infrared spectroscopy. J Cereb Blood Flow Metab 13:516–520
32. Villringer A, Planck A, Hock C, Schleinkofer L, Dirnagl U (1993) Near infrared spectroscopy (NIRS): a new tool to study hemodynamic changes during activation of brain function in human adults. Neurosci Lett 154:101–104
33. Patterson MS, Schwartz E, Wilson BC (1989) Quantitative reflectance spectrophotometry for the noninvasive measurement of photosensitizer concentration in tissue during photodynamic therapy. Proc SPIE 1065:115–122
34. Matcher SJ, Kirkpatrick P, Nahid N, Cope M, Delpy DT (1995) Absolute quantification method in tissue near infrared spectroscopy. Proc SPIE 2389:486–495
35. Suzuki S, Takasaki S, Ozaki T, Kobayashi K (1999) A tissue oxygenation monitor using NIR spatially resolved spectroscopy. Proc SPIE 3597:582–592
36. Al-Rawi PJ, Smielewski P, Kirkpatrick PJ (2001) Evaluation of a near-infrared spectrometer (NIRO 300) for the detection of intracranial oxygenation changes in the adult head. Stroke 32:2492–2500
37. Chance B, Leigh JS, Miyake H, Smiths DS, Nioka S, Greenfeld R, Finander M, Kaufmann K, Levy W, Young M, Cohen P, Yoshioka H, Boretsky R (1988) Comparison of time-resolved and -unresolved measurements of deoxyhemoglobin in brain. Proc Natl Acad Sci USA 85:4971–4975
38. Delpy DT, Cope M, van der Zee P, Arridge S, Wray S, Wyatt JS (1988) Estimation of optical pathlength through tissue from direct time of flight measurement. Phys Med Biol 33(12):1433–1442

39. Nomura M, Hazeki O, Tamura M (1989) Exponential attenuation of light along the nonlinear optical path in the scattered media. Adv Exp Med Biol 248:71–80
40. Patterson MS, Chance B, Wilson BC (1989) Time resolved reflectance and transmittance for the noninvasive measurement of tissue optical properties. Appl Opt 28:2331–2336
41. Oda M, Yamashita Y, Nakano T, Suzuki A, Shimizu K, Hirano I, Shimomura F, Ohmae E, Suzuki T, Tsuchiya Y (2000) Nearinfrared time-resolved spectroscopy system for tissue oxygenation monitor. Proc SPIE 4160:204–210
42. Zhang H, Miwa M, Yamashita Y, Tsuchiya Y (1998) Simple subtraction method for determining the mean path length traveled by photons in turbid media. Jpn J Appl Phys 37–1(2):700–704
43. Ichiji S, Kusaka T, Isobe K, Okubo K, Kawada K, Namba M, Okada H, Nishida T, Imai T, Itoh S (2005) Developmental changes of optical properties in neonates determined by near-infrared time-resolved spectroscopy. Pediatr Res 58(3):568–572
44. Ohmae E, Ouchi Y, Oda M, Suzuki T, Yamashita Y (2006) Cerebral hemodynamics evaluation by near-infrared time-resolved spectroscopy: correlation with simultaneous positron emission tomography measurements. Neuroimage 29:697–705
45. Ohmae E, Oda M, Suzuki T, Yamashita Y, Kakihana Y, Matsunaga A, Kanmura Y, Tamura M (2007) Clinical evaluation of time-resolved spectroscopy by measuring cerebral hemodynamics during cardiopulmonary bypass surgery. J Biomed Opt 12(6):062112
46. Chance B, Hulsizer RI, MacNichol EF Jr, Williams FC (1949) Electronic time measurements, vol 20, MIT Radiation Laboratories Series. Boston Technical, Lexington
47. Ma HY, Du C, Chance B (1997) Homodyne frequency-domain instrument: I&Q Phase detection system. Proc SPIE 2979:826–837
48. Kohl M, Watson R, Cope M (1997) Optical properties of highly scattering media determined from changes in attenuation, phase and modulation depth. Proc SPIE 2979:365–374
49. Feddersen BA, Piston DW, Gratton E (1989) Digital parallel acquisition in frequency domain fluorometry. Rev Sci Instrum 60:2929–2936
50. Madsen SJ, Anderson ER, Haskell RC, Tromberg BJ (1994) Portable, high-bandwidth frequency-domain photon migration instrument for tissue spectroscopy. Opt Lett 19:1934–1936
51. Chance B, Cope M, Gratton E, Ramanujam N, Tromberg B (1998) Phase mesurement of light absorption and scatter in human tissue. Rev Sci Instrum 69:3457–3481
52. Fishkin JB, Gratton E (1993) Propagation of photon-density wave in strongly scattering media containing an absorbing semi-infinite plane bounded by a straight edge. J Opt Soc Am A 10:127–140
53. Fantini S, Franceschini MA, Maier J, Walker S, Barbieri B, Gratton E (1995) Frequency-domain multichannel optical detector for noninvasive tissue spectroscopy and oximetry. Opt Eng 34:32–42

Photon Migration in Tissue

2

Masatsugu Niwayama and Yutaka Yamashita

2.1 Introduction

Although many researchers have attempted to determine the absolute value of tissue oxygenation using time-resolved spectroscopy, spatially resolved spectroscopy (SRS), phase-modulated spectroscopy, and continuous-wave spectroscopy (CWS) (see Chap. 1), correction methods are necessary for quantitative measurement. For example, such overlying tissues as skulls and subcutaneous adipose tissues greatly affect the measurement sensitivity of near-infrared spectroscopy (NIRS). Therefore, analysis of photon migration is important in obtaining accurate absolute measurements. Several researchers have derived equations for the temporal and spatial dependence of diffusely reflected light in a turbid medium [1–8]. Analytical solutions of diffusion theory are widely used to quantify the optical properties of homogeneous media. Kienle et al. [8] theoretically examined propagation in a two-layered medium using a diffusion theory. In addition, Monte Carlo methods have also frequently been used to simulate photon migration [9–16]. In 1983 Wilson et al. [9] presented a Monte Carlo model to examine light propagation in tissues. Van der Zee and Delpy [10] proposed a new method for calculating absorption during light propagation. Many researchers have analyzed various models that simulate actual tissue structure. For example, Wang et al. [11] performed Monte Carlo modeling of photon transport in multilayered tissues. Okada et al. [12] investigated the influence of cerebrospinal fluid and the skull on cerebral oxygenation measurement using a Monte Carlo method. Yamamoto and Niwayama [13, 14] also performed a Monte Carlo simulation of muscle oxygenation measurement with a four-layered model and demonstrated a simple correction method for CWS. Additionally, Boas et al. [15, 16] reported high-speed analysis of three-dimensional photon migration using graphics processing units.

In this chapter we present methods for analyzing photon migration in tissues. The effects of inhomogeneities on NIRS, determined by photon migration analysis are also described.

M. Niwayama, Ph.D. (✉)
Department of Electrical and Electronic Engineering, Shizuoka University, 3-5-1 Johoku, Nakaku, Hamamatsu, Shizuoka 432-8561, Japan
e-mail: tmniway@ipc.shizuoka.ac.jp

Y. Yamashita, BS
Central Research Laboratory, Hamamatsu Photonics KK, 5000 Hirakuchi, Hamakita-ku, Hamamatsu City, Shizuoka 434-8601, Japan
e-mail: yutaka@crl.hpk.co.jp

T. Jue and K. Masuda (eds.), *Application of Near Infrared Spectroscopy in Biomedicine*, Handbook of Modern Biophysics 4, DOI 10.1007/978-1-4614-6252-1_2,

2.2 Photon Diffusion Theory

2.2.1 Homogeneous Media

Exact analytical solutions of the radiation transport equation have been found in only a few special cases [1], and it is a formidable task to perform numerical calculations using the discrete-ordinate method, even for simple geometries. Thus, many investigators have used angle-independent solutions of a transport equation based on diffusion (or P1) approximation to model light scattering in biological tissues. The radiation transport equation is given as

$$\frac{1}{c}\frac{\partial I(\mathbf{r},\mathbf{s},t)}{\partial t} = -\boldsymbol{S}\cdot\nabla I(\boldsymbol{r},\boldsymbol{s},t) - (\mu_s+\mu_a)I(\boldsymbol{r},\boldsymbol{s},t) + \frac{\mu_s}{4\pi}\int_{4\pi} I(\boldsymbol{r},\boldsymbol{s},t)p(\boldsymbol{s},\boldsymbol{s}')ds' + q(\boldsymbol{r},\boldsymbol{s},t), \quad (2.1)$$

where $I(\boldsymbol{r},\boldsymbol{s},t)$ is the average power flux density at point $\boldsymbol{r}$ in the direction of unit vector $\boldsymbol{s}$ (Fig. 2.1), $p(\boldsymbol{s},\boldsymbol{s}')$ is the scattering phase function, and $q(\boldsymbol{r},\boldsymbol{s},t)$ is the source function (the number of photons injected into a unit volume).

In a diffusive process, the following assumptions are made: (1) scattering dominates absorption in the medium, (2) scattering is nearly isotropic, and (3) measurements are made at a sufficient distance from sources and boundaries. Under these assumptions, the diffusion equation can be derived from the radiative transport equation as follows:

$$\frac{1}{c}\frac{\partial\phi(r,t)}{\partial t} - D\nabla^2\phi(r,t) + \mu_a\phi(\boldsymbol{r},t) = S(\boldsymbol{r},t), \quad (2.2)$$

where total energy fluence rate $\phi(\boldsymbol{r},t)$, isotropic source $S(\boldsymbol{r},t)$, and diffusion constant D are given by

$$\phi(\boldsymbol{r},t) = \int_{4\pi} I(\boldsymbol{r},\boldsymbol{s},t)\, ds, \quad (2.3)$$

$$S(\boldsymbol{r},t) = \int_{4\pi} q(\boldsymbol{r},\boldsymbol{s},t)\, ds, \quad (2.4)$$

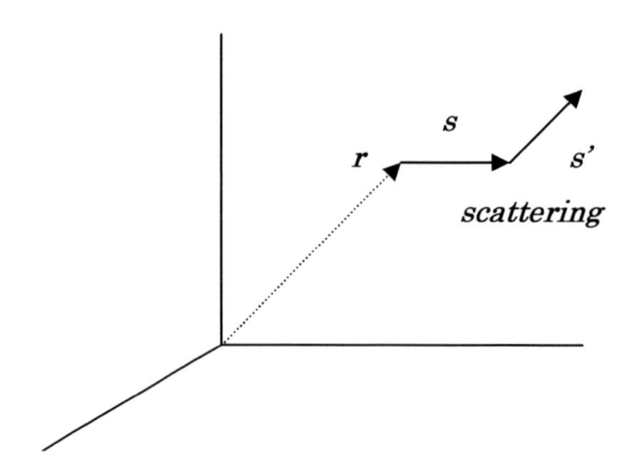

Fig. 2.1 Photon transport

$$D = \frac{1}{3(\mu'_s + \mu_a)}. \tag{2.5}$$

Patterson et al. [4] derived the following equation for the temporal and spatial dependence of diffusely reflected light (reflectance R) in a semiinfinite medium:

$$R(\rho, t) = (4D\pi c)^{-3/2} t^{-5/2} z_0 \exp(-\mu_a ct) \exp\left(-\frac{z_0^2 + \rho^2}{4Dct}\right). \tag{2.6}$$

Integration over a time t yields [6]

$$R(\rho) = \frac{z_0}{2\pi}\left(\mu_{\text{eff}} + \frac{1}{\sqrt{z_0^2 + \rho^2}}\right)\frac{\exp\left(-\mu_{\text{eff}}\sqrt{z_0^2 + \rho^2}\right)}{z_0^2 + \rho^2}. \tag{2.7}$$

If $\rho > z_0$, reflectance R is given by

$$R(\rho) = \frac{1}{2\pi\mu'_s\rho^2}\left(\mu_{\text{eff}} + \frac{1}{\rho}\right)\exp(-\mu_{\text{eff}}\rho), \tag{2.8}$$

2.2.2 Inhomogeneous Media

Several investigators have investigated a solution for a diffusion equation for layered turbid media. Takatani et al. [3] derived analytical expressions for steady-state reflectance by using Green's functions to solve the diffusion equation, and Dayan et al. [5] used Fourier and Laplace transforms to obtain solutions for steady-state and time-resolved reflectances. Kienle et al. [8] solved the diffusion equation using a Fourier transform approach for a two-layered turbid medium as follows:

$$R(\rho) = 0.118\Phi_1(\rho, z = 0) + 0.306\frac{\partial}{\partial z}\Phi_1(\rho, z)|_{z=0}, \tag{2.9}$$

$$\Phi_1(\rho, z) = \frac{1}{2\pi}\int_0^\infty \phi_1(z) s J_0(s\rho)\, ds, \tag{2.10}$$

$$\phi_1(z) = \frac{\sinh(\alpha_1(z_b + z_0))}{D_1\alpha_1}\frac{D_1\alpha_1\cosh(\alpha_1(l - z)) + D_2\alpha_2\sinh(\alpha_1(l - z))}{D_1\alpha_1\cosh(\alpha_1(l + z_b)) + D_2\alpha_2\sinh(\alpha_1(l + z_b))} - \frac{\sinh(\alpha_1(z_0 - z))}{D_1\alpha_1}, \tag{2.11}$$

where J_0 is the zeroth-order Bessel function, D_i is the diffusion constant for layer i, $\alpha_i^2 = (D_i s^2 + \mu_{ai})/D_i$, l is the thickness of the first layer, z_0 is the isotropic source depth, $z_b = (1 + R_{\text{eff}})D_1/(1 - R_{\text{eff}})$, and R_{eff} is 0.493 [7], for a refractive index of 1.4. The reflectance calculated from this solution agreed well with that computed by a Monte Carlo simulation. Although this approach was similar to that proposed by Dayan, it was more accurate and considered mismatches in the refraction index at a tissue surface.

The finite-element method (FEM) is the most common method for numerically solving the diffusion equation for arbitrarily shaped inhomogeneous media. Analysis of light propagation using the FEM [17] is based on the discrete diffusion equation. Furthermore, the model for the FEM is divided into a large number of volume or area elements, each of which has its individual set of optical properties (absorption and scattering coefficients). Solutions are found simultaneously at all nodes of the finite-element mesh by inverting the associated matrix. The photon densities of the individual elements are computed from the node values via an interpolation scheme, which ensures continuity of the overall solution. Although the FEM is fast compared to the Monte Carlo method (see Sect. 2.3), it requires a long time and a large amount of computer memory for calculating a three-dimensional model having a complex shape.

2.3 Monte Carlo Methods

The statistical behavior of random walks has been used to examine light propagation in a turbid medium. Bonner et al. [18], Nossal et al. [19], and Taitelbaum et al. [20] modeled the kinetics of photon migration in a two-layered medium in terms of a random walk on a discrete lattice, where the lattice spacing is equivalent to the root-mean-square distance between scattering events, and absorption occurs in the intervening space. Taitelbaum et al. have found that if the upper layer has a higher absorption than the lower layer, the absorption coefficients of the two layers can be determined from a surface reflectance profile. In contrast, when the upper layer has a lower absorption than the lower layer it is difficult to estimate the optical properties of the two regions.

The Monte Carlo technique has been widely applied to radiation transport studies. This technique is based on the stochastic nature of radiation interactions. The probability of a photon being scattered after traveling a distance dl is defined as

$$p(l)\ dl = \exp(-\mu_s l)\ dl. \tag{2.12}$$

Thus, the cumulative probability of scattering after a distance l is traveled is

$$\int_0^l p(l')\ dl' = 1 - \exp(-\mu_s l) \equiv r, \tag{2.13}$$

where $0 < r < 1$ is a uniformly distributed random number; therefore, $1 - r$ equals r. The path length between scattering events is calculated as

$$l = -\frac{1}{\mu_s}\ln(1-r) = -\frac{1}{\mu_s}\ln(r). \tag{2.14}$$

Wilson et al. developed a Monte Carlo model to examine propagation in tissues [9]. In this model (Fig. 2.2a), at the absorption/scattering point, the photon is assumed to deposit a fraction μ_a/μ_t of its current intensity I_i (initially set equal to 1) as absorbed energy and to emerge from the point with a weighting factor:

$$I_{i+1} = (\mu_s/\mu_t)I_i. \tag{2.15}$$

The direction of the scattered photon is selected from a random distribution such that the probability per unit solid angle is the same in all directions. In contrast, van der Zee and Delpy [10] proposed

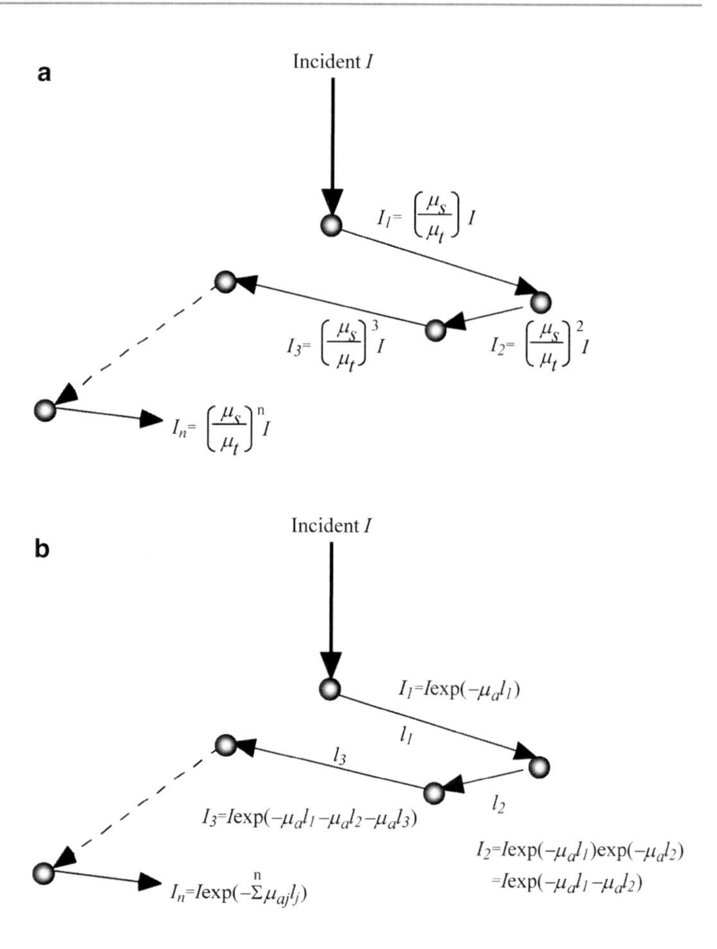

Fig. 2.2 Monte Carlo models reported by Wilson et al. (**a**) and van der Zee and Delpy (**b**)

a new method for calculating absorption in media. In their method absorption occurs at the molecular level and is therefore equal to $\exp(-\mu_a l)$, as shown in Fig. 2.2b. It is shown here that these two methods are essentially the same. From Eq. 2.15, detected light intensity I is expressed as

$$I = (\mu_s/\mu_t)^N I_0 = (1 - \mu_a/\mu_t)^N I_0, \tag{2.16}$$

where I_0 is the initial light intensity and N is the number of scattering events.

The number of scattering events N is equal to $\mu_s' l$. Maclaurin expansion yields the following expression:

$$\begin{aligned}\left(1 - \frac{\mu_a}{\mu_t}\right)^{\mu'_s l} &= 1 - \left(\mu'_s l \frac{\mu_a}{\mu_t}\right) + \frac{\mu'_s l(\mu'_s l - 1)}{2!}\left(\frac{\mu_a}{\mu_t}\right)^2 - \frac{\mu'_s l(\mu'_s l - 1)(\mu'_s l - 2)}{3!}\left(\frac{\mu_a}{\mu_t}\right)^3 \\ &\quad + \cdots + (-1)^n \frac{\mu'_s l(\mu'_s l - 1)\cdots(\mu'_s l - n + 1)}{n!}\left(\frac{\mu_a}{\mu_t}\right)^n + \cdots\end{aligned} \tag{2.17}$$

for $\mu_a/\mu_s < 1$.

In contrast, detected light intensity I in the model proposed by van der Zee and Delpy is expressed as

$$I = I_0 \exp(\mu_a l). \tag{2.18}$$

Maclaurin expansion yields the following:

$$\exp(-\mu_a l) = 1 - (\mu_a l) + \frac{1}{2!}(\mu_a l)^2 - \frac{1}{3!}(\mu_a l)^3 + \cdots \tag{2.19}$$

For all $\mu_a l$.

Hence, Eqs. 2.17 and 2.19 show that the model proposed by van der Zee and Delpy is almost identail to of Wilson et al.

2.4 Models for Monte Carlo Simulation

The three-dimensional models shown in Fig. 2.3 were used to simulate light propagation within a layered structure. A two-layered model (fat and muscle layers) was used initially. To examine the effect of the skin, a four-layered model (epidermis, dermis, fat, and muscle layers) was also employed. The thicknesses of the epidermis and dermis were 60 μm and 1 mm, respectively. The thickness of the fat

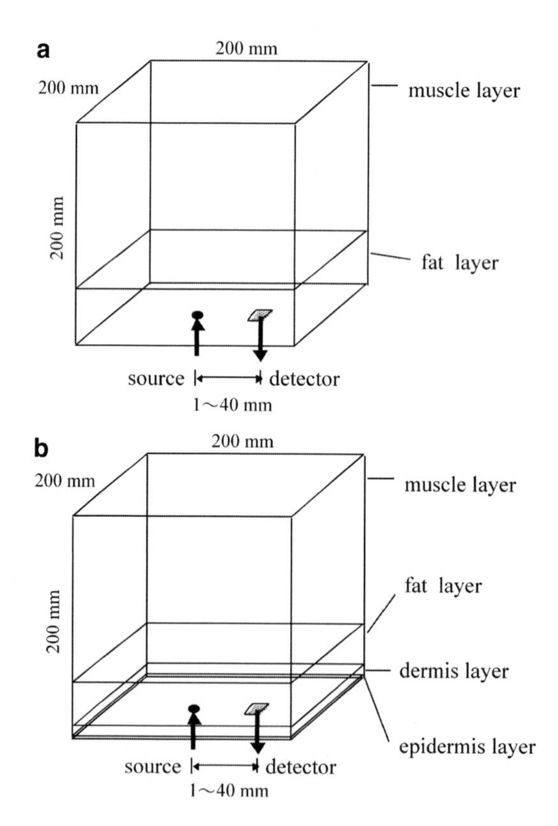

Fig. 2.3 Two- (**a**) and four-layered (**b**) models

Table 2.1 Reduced scattering coefficient and absorption coefficient of each layer

Layer	Reduced scattering coefficient (mm^{-1})	Absorption coefficient (mm^{-1})	References
Epidermis	5.0	5.9	[21]
Dermis	1.3	0.03	[22]
Fat	1.2	0.003	[23]
Muscle	0.6	0.02	[24]

layer varied from 0 to 14 mm, while the total thickness of the four layers was maintained at 200 mm. The size of the model was 200 mm × 200 mm × 200 mm. The source–detector distance varied from 20 to 40 mm.

Calculations were performed by the algorithm [10] presented below. Isotropic scattering length L_i of the ith path was determined by Eq. 2.20, as described in Sect. 2.3, as follows:

$$L_i = -\ln(R)/\mu'_s, \tag{2.20}$$

where R is a random number between 0 and 1 and μ'_s is the reduced scattering coefficient of each layer. When a photon crossed the boundary between the fat and muscle layers, L_i was corrected by the ratio of the reduced scattering coefficients of both layers. Because biological tissues are strong multiple-scattering media, isotropic scattering was assumed. The successive paths of a photon in the two layers were stored, and the relative intensity of a detected photon was calculated by Eq. 2.21, which was derived from Eq. 2.18, as

$$I/I_0 = \exp(-\mu_{af}\ \Sigma L_i - \mu_{am}\ \Sigma L_j) \tag{2.21}$$

where μ_{af} and μ_{am} are the absorption coefficients of the fat and muscle layers, respectively. In the simulation, 10^8 photons were vertically injected into the fat layer at the source point. The size of the detector was assumed to be 5 mm × 4 mm, which corresponds to the size of the photodiode (S2386-45K, Hamamatsu Photonics KK).

The optical properties of the simulated tissues were determined on the basis of values in the literature [21–23]. The reduced scattering coefficients and absorption coefficients of each layer are shown in Table 2.1.

2.5 Calculation of Spatial Sensitivity

Photon migrations within a fat layer and a muscle layer were obtained by Monte Carlo simulation. The spatial sensitivities at different tissue sites, showing the contribution of different tissue sites to the intensity of the detected light, were also calculated. The spatial sensitivity is also equivalent to the mean optical path length of each cell defined at the sites of interest. The model was divided into small cubes, called cube "cells," as shown in Fig. 2.4. Each cube was 1 mm × 1 mm × 1 mm in size. The mean path length $<L>$ of a cell was calculated by $\sum(L_iI_i)/\sum I_i$, where L_i is the path length of the ith photon in the cell and I_i is the final intensity of the ith photon at a detector. To focus on the spatial sensitivities in the x- and z-axis directions, the sensitivity values of the cell were summed up in the y-axis direction and illustrated. The distribution of sensitivities in the x-y plane at each layer was also obtained.

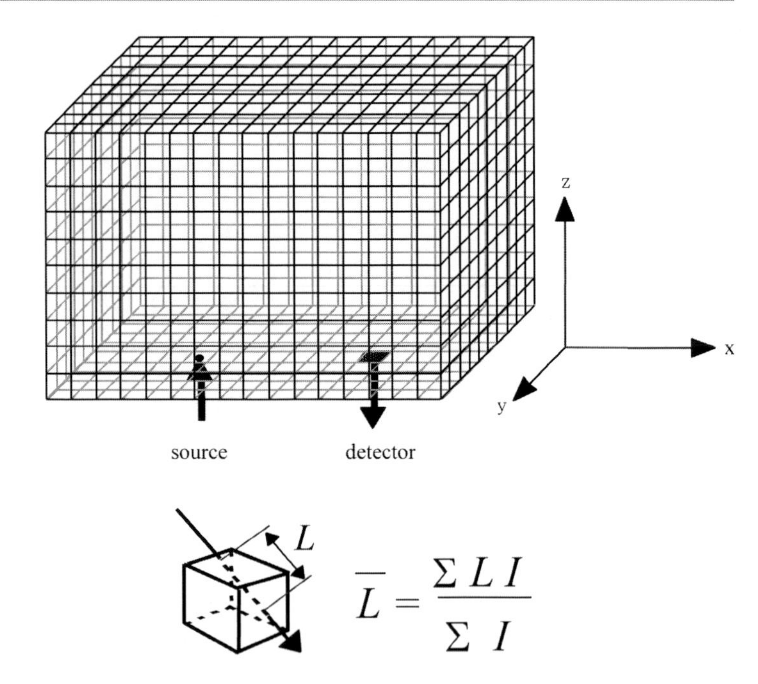

Fig. 2.4 Model divided into cells and calculation of mean optical path length <*L*>

2.6 Photon Migration in Layered Tissues

Figure 2.5 shows light propagation within a fat layer and a muscle layer obtained by simulation using a two-layered model. This graphic shows the distribution of spatial sensitivities (i.e., the mean optical path length of each cell). The sensitivity distribution has a banana shape. Sites close to the source or the detector have the greatest effect on the detected light. When the fat layer is 3 mm thick, the spatial sensitivity within the muscle layer is relatively high. In contrast, the spatial sensitivity of the muscle layer is greatly diminished when the fat layer is 10 mm thick. Because the absorption coefficient of a fat layer is much lower than that of a muscle layer, the detected light consists mainly of light passing through the fat layer when it is thick. Light that penetrates into the muscle and reaches the detector is greatly reduced. This implies that the presence of a fat layer greatly affects measurement sensitivity.

Figure 2.6 shows the sensitivity distribution in the x–y plane at a source–detector separation of 30 mm. The distribution is spindle shaped and is 10 mm wide in the y-axis direction. These results are important for the basic design of an imaging system.

2.7 Effect of Fat Layer

Figure 2.7 shows the mean optical path lengths $<L_m>$ in a muscle layer at source–detector separations of 20, 30, and 40 mm, which were obtained by Monte Carlo simulation. The presence of a fat layer greatly decreases $<L_m>$. When the source–detector separation is small, $<L_m>$ becomes zero even for a relatively thin fat layer.

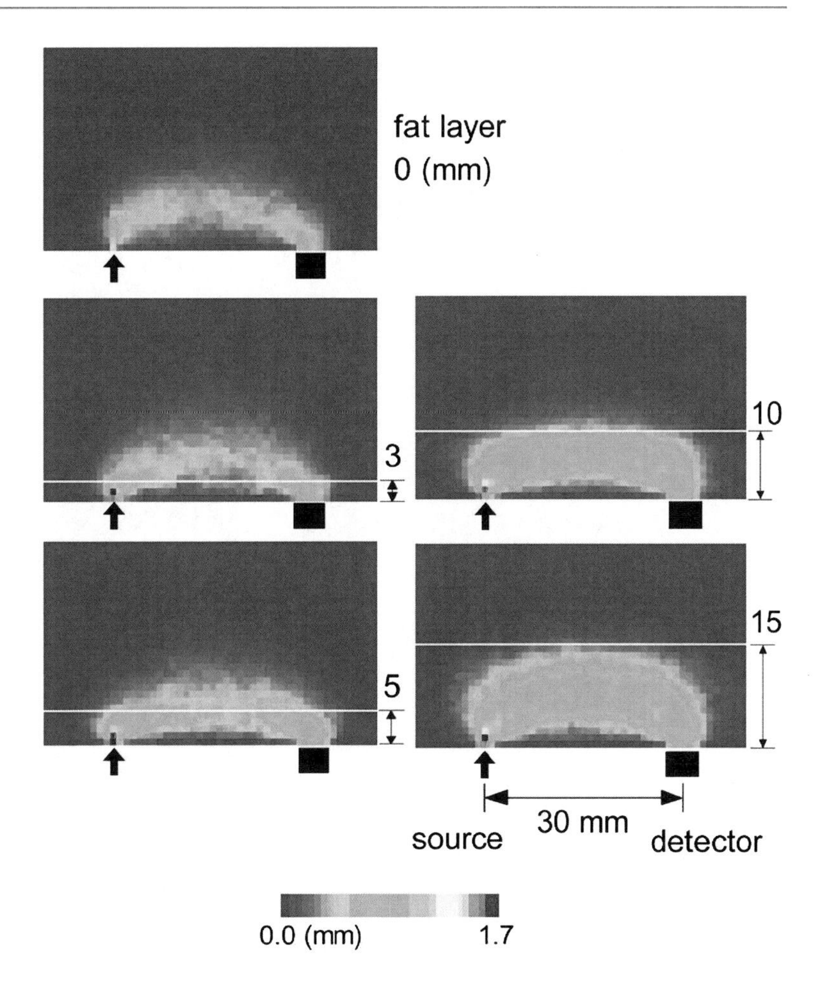

Fig. 2.5 Spatial sensitivities for various fat layer thicknesses

Figure 2.8 shows the optical path length in a muscle layer obtained from a Monte Carlo simulation in which the absorption coefficient for the muscle was 0.02 mm^{-1}. The relationship between normalized optical path length S_{muscle} and fat layer thickness h is expressed by the following equation:

$$S_{muscle} = \exp\left\{-\left(\frac{h}{A}\right)^2\right\}. \tag{2.22}$$

The constant A has the values 6.9, 8.0, and 8.9 for source–detector distances of 20, 30, and 40 mm, respectively. The value of S_{muscle} can be determined only by using the h value previously measured by ultrasonography. Then, the changes in hemoglobin concentration obtained using CW-NIRS can be corrected by dividing them by S_{muscle}. In in vivo tests the average value of resting oxygen consumption in the forearms of the subjects was 0.13 ± 0.02 ml 100 g^{-1} min^{-1} (mean $\pm$ SD) after correction. Before correction of a fat layer, the mean value of oxygen consumption was about 50% smaller than the reported values [25, 26], and the coefficient of variation showed great dispersion.

A decrease in the measurement sensitivity is caused by an increase in the proportion of light that directly reaches the detector through the fat layer. Therefore, not only the measurement sensitivity but also the intensity of detected light depends on fat layer thickness. Therefore, the detected light intensity was recorded and plotted against fat layer thickness, as depicted in Fig. 2.9. Detected light intensity increased with fat layer thickness. Light intensity is plotted on a logarithmic scale. When the fat layer is

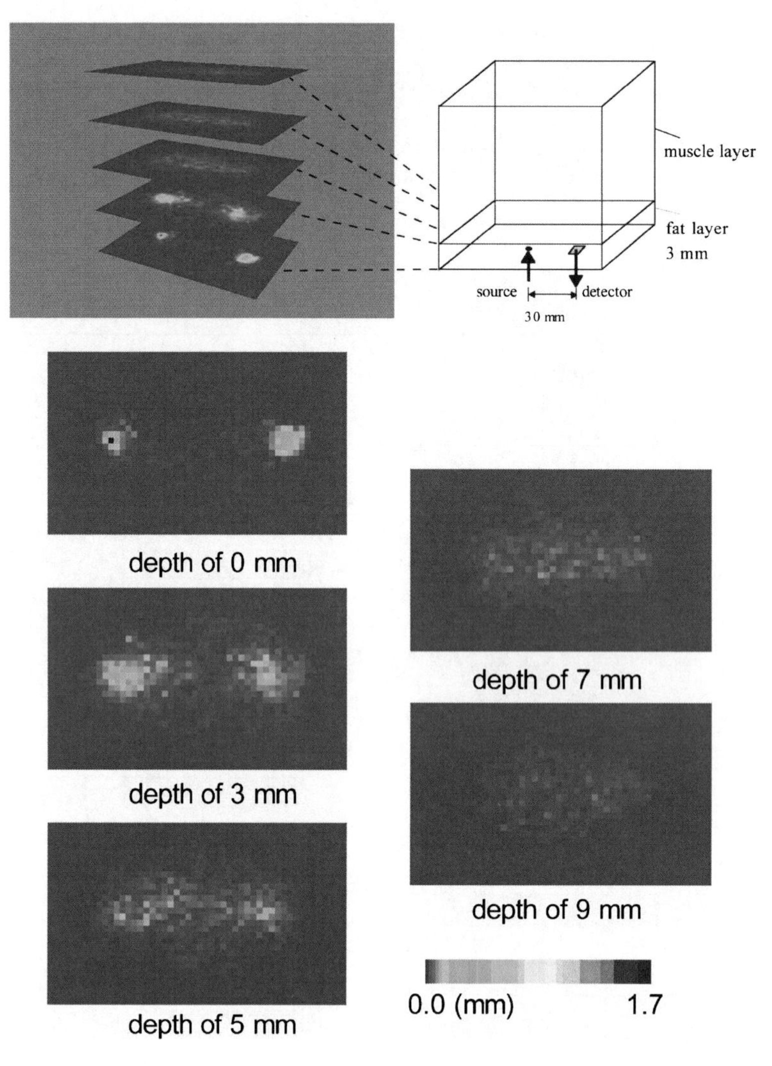

Fig. 2.6 Spatial sensitivities at depths of 0–9 mm. Fat layer thickness, 3 mm; source–detector separation, 30 mm

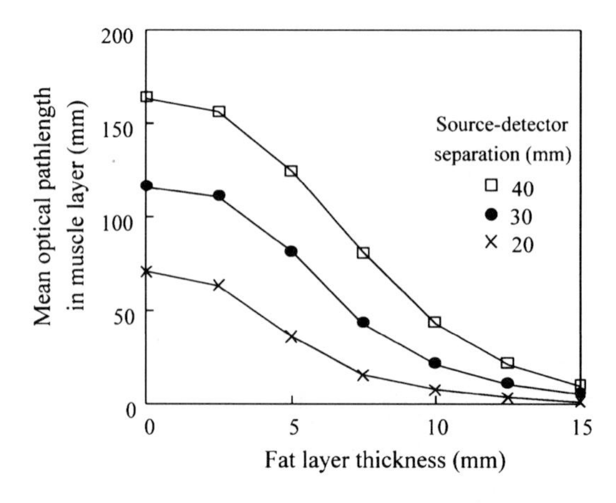

Fig. 2.7 Decrease in measurement sensitivity because of presence of a fat layer. Mean optical path length in the muscle layer was calculated by Monte Carlo simulation

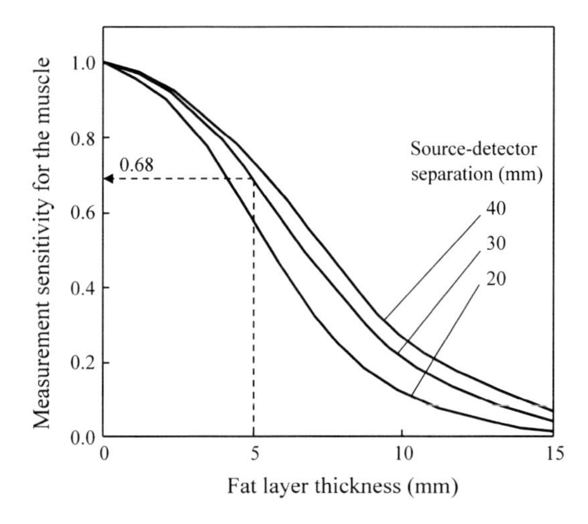

Fig. 2.8 Measurement sensitivity for muscle in CW-NIRS calculated by Monte Carlo simulation

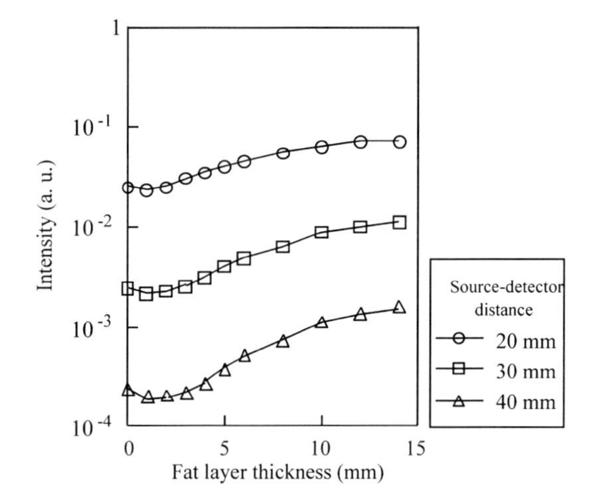

Fig. 2.9 Relationship between detected light intensity and fat layer thickness in the simulation results

thicker than 9 mm, detected light intensity increases slightly and reaches a maximum, corresponding to the light intensity detected in a fat layer whose thickness is regarded as infinite. Therefore, when the source–detector separation is small, the maximum intensity is reached at a small fat layer thickness. For a fat layer thickness of less than 3 mm, an upward tendency appears around a fat layer thickness of zero. This is thought to be due to an increase in light propagating into a muscle layer and light backscattered from this layer to a detector, because attenuation due to scattering in a fat layer diminishes with decreasing fat layer thickness.

2.8 Effect of Skin

The effect of the skin on muscle oxygenation measurement was also examined by simulation. Figure 2.10a shows the relationship between the mean optical path length in a muscle layer and the thickness of a fat layer at various source–detector distances. When the source–detector distance is less

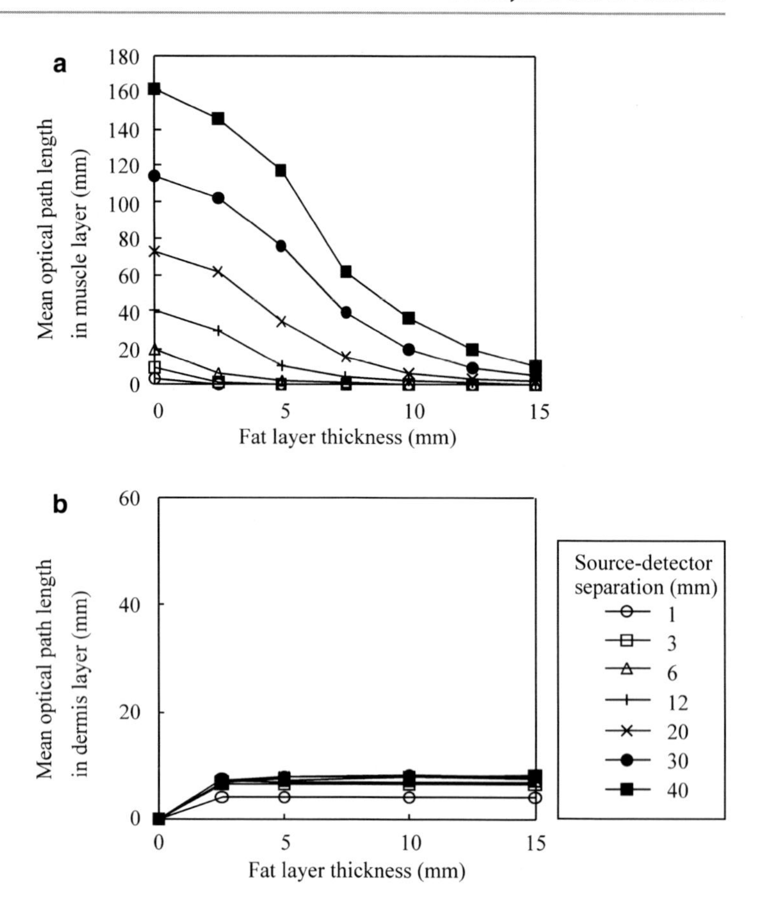

Fig. 2.10 Relationship between fat layer thickness and mean optical path length in a muscle layer (**a**) and in a dermis layer (**b**)

than 3 mm and the fat layer is thicker than 3 mm, the mean optical path length in the muscle layer is almost zero. Thus, there is almost no sensitivity toward changes in muscle oxygenation under this condition. The mean optical path length in the dermis is almost constant when the source–detector distance and fat layer thickness are larger than 3 mm, as shown in Fig. 2.10b. These results suggest that the effect of skin blood flow can be eliminated by using the signal obtained at a 3-mm separation. Subtraction of optical density (OD) at the 3-mm separation from that at larger separations yields the change in muscle oxygenation without the effect of the skin and without decreasing sensitivity to the muscle. Thus, the optimal source–detector separation for eliminating the effect of the skin is likely to be 3 mm.

However, the effect of the skin on change in OD was 8% or less compared to that of muscle at the 30-mm separation, and there was little change in blood volume of the skin because the optical probe was pressed lightly on the body surface in in vivo measurements. Thus, the effect of blood in the skin can be ignored when the source–detector separation is large (>20 mm).

2.9 Analysis of Time-Resolved Measurement

Diffuse optical tomography using time-resolved reflectance has been widely studied as a useful technique for noninvasive measurement of internal physiological information. Analysis of a time-resolved photon path distribution is essential for conducting image reconstruction using optical tomography. Figure 2.11 shows the impulse response without absorption and the photon path distributions

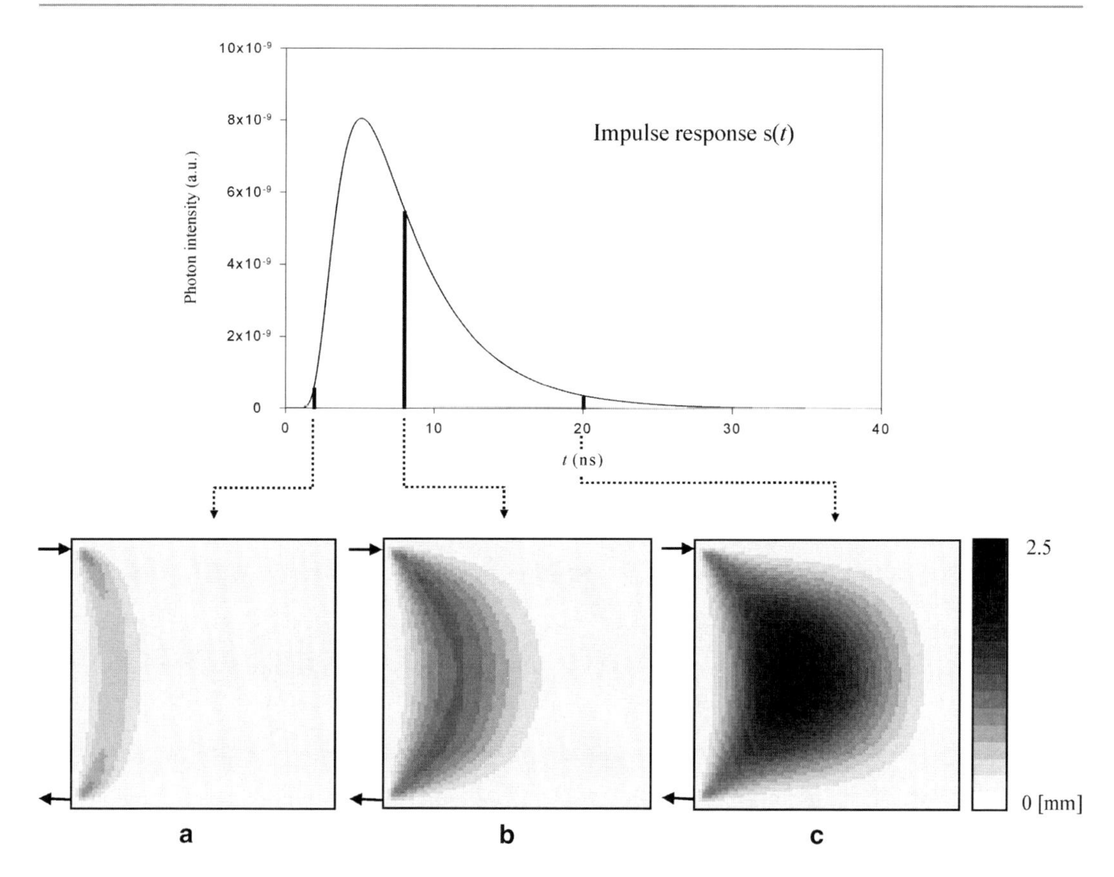

Fig. 2.11 Impulse response and photon path distributions at $t = 2$ ns (**a**), $t = 8$ ns (**b**), and $t = 20$ ns (**c**)

at 2, 8, and 20 ns, calculated using diffusion theory [27]. A two-dimensional finite-volume method was used for the analysis. The grid size was 1 mm, and the time interval was 2 ps. The path distribution of photons that arrived at 2 ns is narrow and restricted to superficial media, whereas the photons detected at 20 ns are sensitive to deep layers. These results allow us to calculate weight functions for various diffuse optical tomography applications, which will be used in future studies.

2.10 Analysis of Spatially Resolved Measurement

To acquire quantitative measurements of the hemoglobin concentration in deep tissues by spatially resolved measurements, it is necessary to analyze the spatial intensity profile for a complex structure [28]. Figure 2.12 shows the increase in spatial slope because of increased absorption coefficient of muscle (μ_{am}) and increased fat layer thickness calculated in the Monte Carlo simulation. The model consisted of skin, fat, and muscle layers; the number of photons was 10^7. The spatial slope S was calculated by $\ln(I_{20}/I_{30})/(30–20)$, where I_{20} and I_{30} are the light intensities at 20 and 30 mm from the light source, respectively. It was found that fat thickness greatly affects determination of μ_{am}. For example, oxygenation measurements for a 3-mm fat layer decreased 30% if the $S - \mu_{am}$ curve of 9-mm fat thickness was used for quantification. The measurements were successfully corrected using the appropriate $S - \mu_{am}$ curve for fat thickness. The influence of millimeter inhomogeneity (e.g., layered structure) can be corrected using a combination of a diagnostic ultrasound

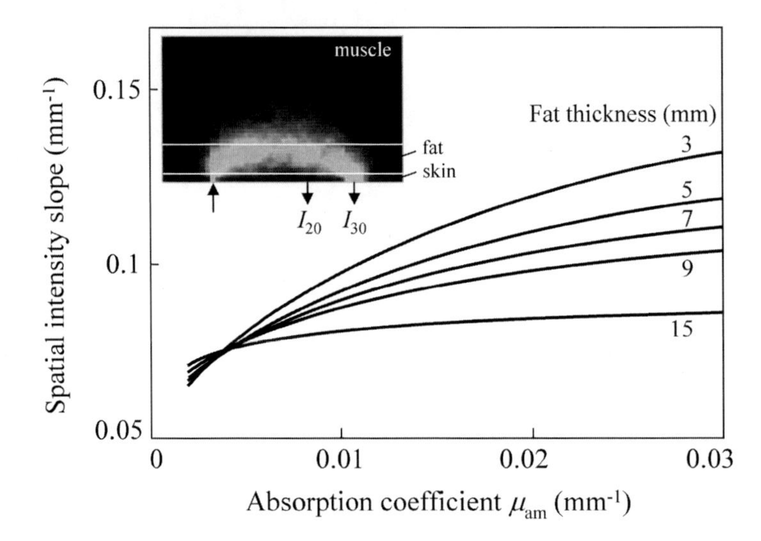

Fig. 2.12 Effect of fat thickness on the relationship between the slope of light intensity and the absorption coefficient of muscle

apparatus and NIRS instrument. Since the effects of micrometer inhomogeneity are reflected in the value of the scattering coefficient, estimation of the optical properties of each layer will be important for accurate oxygenation measurement.

Problem

2.1 A random number used on Monte Carlo simulation should be long period and of almost uniform distribution. How can this random number be generated?

Further Reading

Matsumoto M, Nishimura T (1998) Mersenne twister: a 623-dimensionally equidistributed uniform pseudorandom number generator. ACM Trans Model Comp Sim 8(1):3–30

Panneton F, L'Ecuyer P, Matsumoto M (2006) Improved long-period generators based on linear recurrences modulo 2. ACM Trans Math Softw 32:1–16

References

1. Chandraseklar S (1960) Radiative transfer. Dover, New York
2. Ishimaru A (1978) Diffusion of a pulse in densely distributed scatterers. J Opt Soc Am 68:1045–1050
3. Takatani S, Graham MD (1979) Theoretical analysis of diffuse reflectance from a two-layer tissue model. IEEE Trans Biomed Eng BME26:656–664
4. Patterson MS, Chance B, Wilson BC (1989) Time-resolved reflectance and transmittance for the noninvasive measurement of tissue optical properties. Appl Opt 28:2331–2336
5. Dayan I, Havlin S, Weiss GH (1992) Photon migration in a two-layer turbid media: a diffusion analysis. J Mod Opt 39:1567–1582
6. Farrell TJ, Patterson MS (1992) A diffusion theory model of spatially resolved, steady-state diffuse reflectance for the noninvasive determination of tissue optical properties in vivo. Med Phys 19:879–888
7. Haskell RC, Svaasand LO, Tsay TT, Feng TC, McAdams M, Tromberg BJ (1994) Boundary conditions for the diffusion equation in radiative transfer. J Opt Soc Am A 11:2727–2741

8. Kienle A, Patterson MS, Dögnitz N, Bays R, Wagnières G, van den Bergh H (1998) Noninvasive determination of the optical properties of two-layered turbid media. Appl Opt 37:779–791
9. Wilson BC, Adam G (1983) A Monte Carlo model for the absorption and flux distributions of light in tissue. Med Phys 10:824–830
10. van der Zee P, Delpy DT (1987) Simulation of the point spread function for light in tissue by a Monte Carlo method. Adv Exp Med Biol 215:179–191
11. Okada E, Firbank M, Delpy DT (1995) The effect of overlying tissue on the spatial sensitivity profile of near-infrared spectroscopy. Phys Med Biol 40:2093–2108
12. Wang L, Jacques S, Zheng L (1995) MCML—Monte Carlo modeling of light transport in multi-layered tissues. Comput Methods Programs Biomed 47(2):131–146
13. Yamamoto K, Niwayama M, Shiga T, Lin L, Kudo N, Takahashi M (1998) Accurate NIRS measurement of muscle oxygenation by correcting the influence of a subcutaneous fat layer. Proc SPIE 3194:166–173
14. Niwayama M, Lin L, Shao J, Kudo N, Yamamoto K (2000) Quantitative measurement of muscle hemoglobin oxygenation using near-infrared spectroscopy with correction for the influence of a subcutaneous fat layer. Rev Sci Instrum 71(12):4571–4575
15. Fang Q, Boas DA (2009) Monte Carlo simulation of photon migration in 3D turbid media accelerated by graphics processing units. Opt Express 17:20178–20190
16. Alerstam E, Svensson T, Andersson-Engels S (2008) Parallel computing with graphics processing units for high-speed Monte Carlo simulation of photon migration. J Biomed Opt 13:060504
17. Arridge SR, Schweiger M, Hiraoka M, Delpy DT (1993) A finite-element approach for modeling photon transport in tissue. Med Phys 20:299–309
18. Bonner R, Nossal R, Havlin S, Weiss H (1987) Model for photon migration in turbid biological media. J Opt Soc Am A 4:423–432
19. Nossal R, Kiefer J, Weiss GH, Bonner R, Taitelbaum H, Havlin S (1988) Photon migration in layered media. Appl Opt 27:3382–3391
20. Taitelbaum H, Havlin S, Weiss GH (1989) Approximate theory of photon migration in a two-layer medium. Appl Opt 28:2245–2249
21. Wan S, Anderson RR, Parrish JA (1981) Analytical modeling for the optical properties of skin with in vitro and in vivo applications. Photochem Photobiol 34:493–499
22. Beek JF, van Staveren HJ, Posthumus P, Sterenborg HJ, van Gemert MJ (1993) The influence of respiration on optical properties of piglet lung at 632.8 nm. In: Medical optical tomography. SPIE Optical Engineering Press, Bellingham, pp 193–210
23. Mitic G, Közer J, Otto J, Plies E, Sökner G, Zinth W (1994) Time-gated transillumination of biological tissues and tissuelike phantoms. Appl Opt 33:6699–6710
24. Zaccanti G, Taddeucci A, Barilli M, Bruscaglioni P, Martelli F (1995) Optical properties of biological tissues. Proc SPIE 2389:513–521
25. Wang ZY, Noyszewski EA, Leigh JS Jr (1990) In vivo MRS measurement of deoxymyoglobin in human forearms. Magn Reson Med 14:562–567
26. Harris RC, Hultman E, Kaijser L, Nordesjö LO (1975) The effect of circulatory occlusion on isometric exercise capacity and energy metabolism of the quadriceps muscle in man. Scand J Clin Lab Invest 35:87–95
27. Ueda Y, Ohta K, Yamashita Y, Tsuchiya Y (2003) Calculation of photon path distribution based on photon behavior analysis in a scattering medium. Opt Rev 10:444–446
28. Niwayama M, Sone S, Murata H, Yoshida H, Shinohara S (2007) Errors in muscle oxygenation measurement using spatially-resolved NIRS and its correction. J Jpn Coll Angiol 47:17–20

3 Photon Migration in NIRS Brain Imaging

Eiji Okada

3.1 Introduction

Near-infrared spectroscopy (NIRS) is widely used to measure cerebral oxygenation and hemodynamics caused by brain activation. Blood volume and oxygenation are indicated by the absorption of tissue caused by oxygenated and deoxygenated hemoglobin. NIRS instruments can monitor temporal changes in blood volume and oxygenation in a single probing region [1–5]. The desire to measure the spatial distribution of tissue absorption, which indicates the region of focal brain activation, has fostered development of NIRS imaging. It is reasonable to design NIRS imaging systems with multiple emission and detection probes to localize the position of an absorption change in the brain. There are two basic categories of NIRS imaging: tomography and topography [6]. NIRS tomography can provide the cross-sectional images of brain activation, whereas the two-dimensional distribution of brain activation in the cortex is obtained by NIRS topography.

The fundamental and most serious problem of NIRS imaging is the ambiguity of light propagation in the brain. Light that propagates in the head is strongly scattered, and the detected light travels significantly farther through tissue than the distance between the emission and detection probes [7]. Information about the distribution of optical path between the emission and detection probes is very important for NIRS imaging; therefore, an adequate model of photon migration in the head has been investigated [8–10]. The optical heterogeneity of superficial tissues, especially the low-scattering cerebrospinal fluid (CSF) layer, affects light propagation in the brain [11–16]. Optical heterogeneity should also be considered when modeling photon migration in the head.

Such a model is described by the geometry and optical properties of each type of tissue when considering photon migration in the head. Layered models and more realistic and efficient models have been developed to make accurate predictions. The anatomical geometry obtained by other modalities (e.g., magnetic resonance imaging (MRI)), is increasingly incorporated into these models [13, 17–21].

Mathematical models of photon migration in scattering tissues employ one of two methods: stochastic and deterministic. Monte Carlo simulation represents a stochastic method, whereas approaches based on solutions of the diffusion equation are the most common deterministic ones [22–29]. A diffusion equation is an approximation of the radiative transport equation and is valid in

E. Okada, Ph.D. (✉)
Department of Electronics and Electrical Engineering, Keio University, Yokohama 223-8522, Japan
e-mail: okada@elec.keio.ac.jp

T. Jue and K. Masuda (eds.), *Application of Near Infrared Spectroscopy in Biomedicine*,
Handbook of Modern Biophysics 4, DOI 10.1007/978-1-4614-6252-1_3,

highly scattering medium. The presence of a low-scattering CSF layer may affect the accuracy of photon migration in the head calculated using the diffusion equation. An adequate model needs to be designed to analyze photon migration with a low-scattering region using the diffusion equation [30–35].

3.2 Important Physical Parameters for NIRS

3.2.1 Modified Beer-Lambert Law

For the measurement of brain activation by common NIRS instruments, a pair of emission-detector probes is attached to the scalp, as shown in Fig. 3.1. The natural logarithm of light attenuation (optical density: OD) across the head measured with a probe pair can be expressed as a modified Beer-Lambert law [7]:

$$\mathrm{OD} = \ln\left(\frac{I_e}{I_d}\right) \approx \mu_{a\,\mathrm{head}}\langle L_{\mathrm{head}}\rangle + G. \tag{3.1}$$

where I_e and I_d are the intensities of the incident and detected light, respectively; $\mu_{a\,\mathrm{head}}$ is the apparent absorption coefficient in the head based on an assumption that absorption in the head is homogeneous; $<L_{\mathrm{head}}>$ is the mean optical pathlength of the detected light; and G is the scattering loss. Absorption in the head cannot be obtained from the optical density because scattering loss G cannot be measured. In typical NIRS measurements, a change in optical density caused by brain activation is used as the NIRS signal, based on an assumption that the scattering loss does not change during the measurement period:

$$\Delta\mathrm{OD} = \ln\left(\frac{I_e}{I'_d}\right) - \ln\left(\frac{I_e}{I_d}\right) = \ln\left(\frac{I_d}{I'_d}\right) = \Delta\mu_{a\,\mathrm{head}}\langle L_{\mathrm{head}}\rangle, \tag{3.2}$$

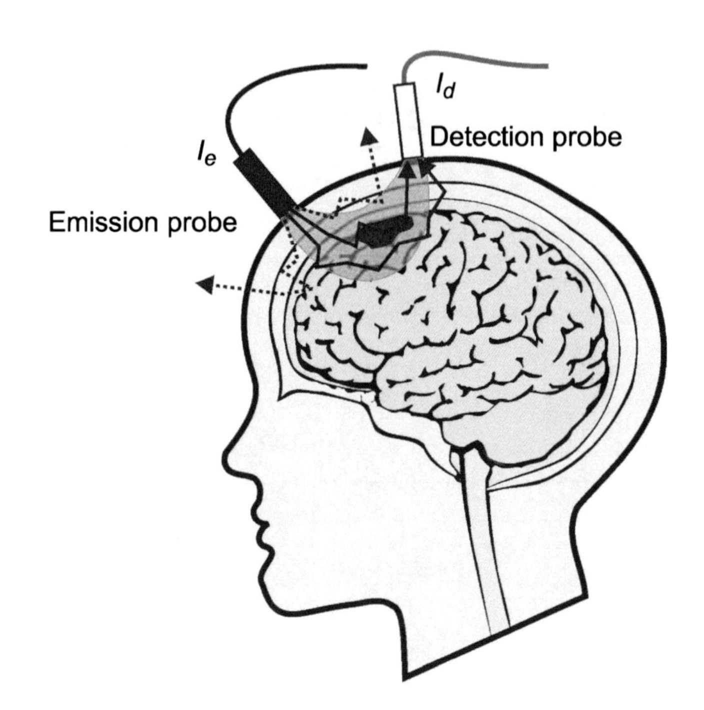

Fig. 3.1 Principle of brain-function measurement by NIRS: Absorption of the activated region in the brain varies by changes in blood volume and oxygenation evoked by brain activation. An emission probe is used for illumination and a detector probe is attached nearby to collect light passing through the head. The change in light intensity caused by absorption change in the brain is detected by the probe pair

where I'_d is the detected intensity in the activation state, and $\Delta\mu_{a\ \mathrm{head}}$ is the apparent absorption change in the head caused by brain activation. The mean optical pathlength in the head can be measured by time- or phase-resolved instrumentation. The apparent absorption change caused by brain activation can be determined from the NIRS signal.

Although tissue contains many absorbing materials, the change in absorption can be described as a change in the concentrations of oxygenated and deoxygenated hemoglobin, based on an assumption that the concentration of hemoglobin is only changed by brain activation:

$$\Delta\mu_{a\ \mathrm{head}}(\lambda) = \varepsilon_{\mathrm{oxy-Hb}}(\lambda)\Delta c_{\mathrm{oxy-Hb}} + \varepsilon_{\mathrm{deoxy-Hb}}(\lambda)\Delta c_{\mathrm{deoxy-Hb}} \tag{3.3}$$

where $\varepsilon_{\mathrm{oxy-Hb}}$ and $\varepsilon_{\mathrm{deoxy-Hb}}$ are the molar extinction coefficients of the oxygenated and deoxygenated hemoglobin, and $\Delta c_{\mathrm{oxy-Hb}}$ and $\Delta c_{\mathrm{deoxy-Hb}}$ are the changes in concentrations of the oxygenated hemoglobin and deoxygenated hemoglobin. The apparent concentration changes in the oxygenated and deoxygenated hemoglobin can be calculated from the change in optical density at two or more wavelengths using Eqs. 3.2 and 3.3. However, an assumption of a homogeneous head for Eq. 3.2 is clearly not realistic.

3.2.2 Partial Optical Pathlength

Absorption in the head is actually changed only locally by brain activation. Equation 3.2 is not strictly applicable for brain function measurements. Assuming that the head consists of several homogeneous tissues, a partial optical pathlength is defined as the mean optical pathlength that the detected light travels in each tissue region. In the case where the head is segmented into M regions, the NIRS signal can be obtained from the change in absorption ($\Delta\mu_{ai}$) and partial optical pathlength ($< L_i >$) in region i as follows:

$$\Delta\mathrm{OD} = \sum_{i=1}^{M} \Delta\mu_{a\,i}\langle L_i\rangle. \tag{3.4}$$

The partial optical pathlength approximates a partial derivative of the change in optical density versus the absorption coefficient of each tissue [8]:

$$\frac{\partial\mathrm{OD}}{\partial\mu_{a\,i}} = \langle L_i\rangle. \tag{3.5}$$

The partial optical pathlength indicates the sensitivity of an NIRS signal to absorption changes in a particular area of the tissue. Although the mean optical pathlength of the head can be measured by time- or phase-resolved experiments, a partial optical pathlength cannot be directly measured by experiment.

A change in absorption in the brain can be obtained from the change in optical density if the absorption change caused by brain activation only occurs in the brain tissue and the activated region is larger than the volume of tissue interrogated by the probe pair. However, the region of brain activation is often smaller than the volume of tissue interrogated; hence, the detected light travels through the activated region that contributes to the NIRS signal which is shorter than the partial optical pathlength in the brain.

3.2.3 Spatial Sensitivity Profile

A more rigorous parameter than partial optical pathlength is necessary to represent the spatial distribution of tissue volume that contributes to the NIRS signal. The volume of tissue in which absorption change contributes to a change in optical density can be described by a spatial sensitivity profile. The spatial sensitivity profile at position $\boldsymbol{r}$, SSP($\boldsymbol{r}$), can be defined as the distribution of the partial optical pathlength in small voxels $<L_{voxel}(\boldsymbol{r})>$ at position $\boldsymbol{r}$ in the head. The partial optical pathlength within a voxel approximates a partial derivative of the change in optical density versus absorption coefficient in the voxel [16, 36, 37].

$$\mathrm{SSP}(\mathbf{r}) = \langle L_{\mathrm{voxel}}(\mathbf{r})\rangle \approx \frac{\partial \mathrm{OD}}{\partial \mu_{a\,\mathrm{voxel}}(\mathbf{r})}. \tag{3.6}$$

Like the partial optical pathlength, the spatial sensitivity profile cannot be directly measured by experiment.

3.3 Brain-Function Imaging by NIRS

3.3.1 Tomography and Topography

In brain-function imaging by NIRS, the absorption change in brain tissue is detected by multiple emission and detection probes attached to the scalp in order to localize functional brain activation. There are two basic types of imaging: tomography and topography.

In NIRS tomography, the emission and detection probes are normally attached around the head as shown in Fig. 3.2a to detect transmitted as well as reflected light. The cross-sectional images of brain function are reconstructed from the NIRS signals detected at various probe spacings [38–40]. Image reconstruction of the cross-sectional image requires solving the inverse problem [41–46]. During x-ray tomography, x-rays propagate along straight lines connecting the radiation sources and detectors, and the forward problem can be described by a Radon transform. During NIRS tomography, light is highly scattered by tissue; hence the detected light propagates through various paths between the emission and detector probes. Since light propagation depends on not only probe positions but also on the optical properties of tissue, a model of photon migration in tissue is essential to predict the spatial sensitivity profiles for image reconstruction of NIRS tomography. In the case of adult heads, it is difficult to detect the transmitted light propagating through the deeper region of the brain because attenuation of light detected with a large probe spacing is extremely high. This fact indicates that the depth of the cross-sectional image of the adult brain is limited. In the case of the neonatal head, an adequate NIRS signal can be detected from the transmitted light. Full three-dimensional images of the neonatal brain can be reconstructed by NIRS tomography [47, 48].

During NIRS topography, the emission and detection probes are two dimensionally arranged to cover an area of the brain surface as shown in Fig. 3.2b in order to detect reflected light from the brain cortex. The distribution of brain activation in the cortex is projected onto a two-dimensional plane on the brain surface [49, 50]. NIRS topography has the advantage of short image acquisition time compared to NIRS tomography because the typical probe spacing for the NIRS topography is about 30 mm. In addition, the forward problem is not essential for NIRS topography because the image can be generated by the mapping method without solving the inverse problem. Based on these advantages, NIRS topography has been extensively applied to brain-function imaging for

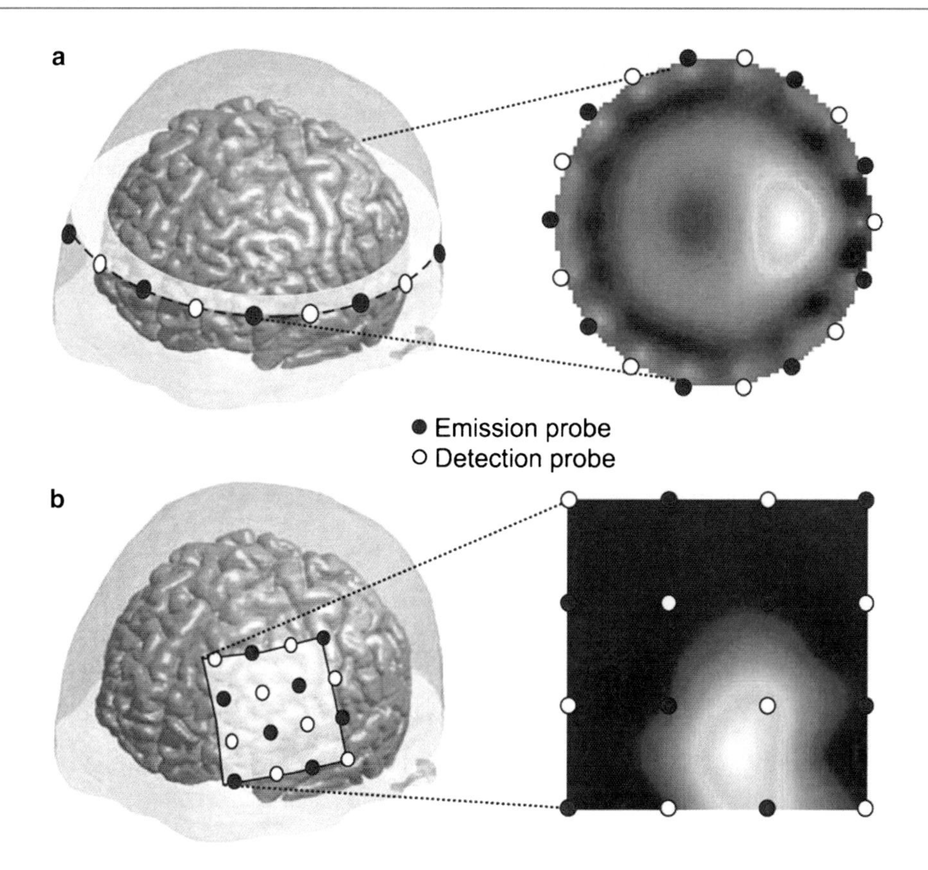

Fig. 3.2 The difference between NIRS topography and NIRS tomography: Schematic diagrams of probe arrangements and generated images of (**a**) NIRS tomography and (**b**) NIRS topography. During NIRS tomography, the emission and detection probes are attached around the head and the cross-sectional image of the brain function is reconstructed from NIRS signals detected at various probe spacings. For NIRS topography, the probes are two dimensionally arranged on the scalp to obtain the distribution of brain activation on the cortex

both adults and neonates [51–54]. However, the spatial resolution and localization accuracy of present-day NIRS topography is not considered sufficient and can be improved by image reconstruction using spatial sensitivity profiles and by employing optimal probe arrangement [55–57].

3.3.2 Reconstruction Method

Image reconstruction method based upon solution of the inverse problem is essential for NIRS tomography. The inverse problem using an appropriate model of photon migration in the head is solved in order to obtain the distribution of absorption change due to brain activation. The forward model for imaging of absorption change can be described as a linear system:

$$\mathbf{y} = \mathbf{J}\,\mathbf{x}, \tag{3.7}$$

where $\boldsymbol{y}$ is the observed data, $\boldsymbol{x}$ are the model parameters, and $\boldsymbol{J}$ is the sensitivity matrix. For brain-function imaging by NIRS, the observed data are the NIRS signals detected by each probe pair, [ΔOD ($\mathbf{r}$)], where $\boldsymbol{\xi}$ represents the positions of the probe pairs; the model parameter is absorption change [$\Delta\mu_a(\mathbf{r})$] at position $\mathbf{r}$ in the image:

$$\begin{bmatrix} \Delta\mathrm{OD}(\xi_1) \\ \vdots \\ \Delta\mathrm{OD}(\xi_i) \\ \vdots \\ \Delta\mathrm{OD}(\xi_M) \end{bmatrix} = \begin{bmatrix} \mathrm{SSP}(\xi_1, r_1) & \cdots & \mathrm{SSP}(\xi_1, r_j) & \cdots & \mathrm{SSP}(\xi_1, r_N) \\ \vdots & \ddots & & & \vdots \\ \mathrm{SSP}(\xi_i, r_1) & & \mathrm{SSP}(\xi_i, r_j) & & \mathrm{SSP}(\xi_i, r_N) \\ \vdots & & & \ddots & \vdots \\ \mathrm{SSP}(\xi_M, r_1) & \cdots & \mathrm{SSP}(\xi_M, r_j) & \cdots & \mathrm{SSP}(\xi_M, r_N) \end{bmatrix} \begin{bmatrix} \Delta\mu_a(r_1) \\ \vdots \\ \vdots \\ \Delta\mu_a(r_j) \\ \vdots \\ \Delta\mu_a(r_N) \end{bmatrix}, \tag{3.8}$$

where M and N are the number of probe pairs and pixels in the image, respectively.

It is necessary to solve the inverse problem to reconstruct the image of brain activation:

$$\mathbf{x} = \mathbf{J}^{-1}\,\mathbf{y}. \tag{3.9}$$

For brain-function imaging by NIRS, number of pixels in image N is much greater than number of probe pairs M. This inverse problem is a non-unique, ill-posed, and underdetermined problem, so that hence regularization is required to stabilize the inverse solution. There are many ways to approach solution of the inverse problem [37, 44, 56, 58]. Prior knowledge, such as the anatomical geometry of the head, is incorporated into the inverse problem to improve reconstructed images [59, 60].

3.3.3 Mapping Method

The mapping method is a simple technique to generate a topographic image of brain activation from NIRS signals. The NIRS signal detected by each probe pair is assumed to be caused by absorption change at the measurement point, which is the midpoint of the emission and detector probes. The sensitivity of the probe pair is normally greatest at the measurement point. The amplitude of the NIRS signal is plotted at the pixel corresponding to the measurement point of the signal. The mapped NIRS signals are connected by spatial interpolation in order to generate the topographic image of brain activation [49, 50].

The advantage of the mapping method is that no prior knowledge, such as the spatial sensitivity profile, is necessary to generate a brain activation image. However, the mapping method cannot be adapted to NIRS tomography, and simple interpolation may reduce the spatial resolution of the topographic image. The activated region in the topographic image generated by the mapping method tends to be broader than actual brain activation [55, 56].

3.4 Head Models for Photon-Migration Analysis

3.4.1 Optical Heterogeneity of Head

Adequate modeling of photon migration in the head to estimate partial optical pathlength in the brain and the spatial sensitivity profile is important for brain-function imaging by NIRS. The head consists of various tissue components that have different optical properties, and hence it is important to investigate the influence of the optical heterogeneity of the head on light propagation in the brain.

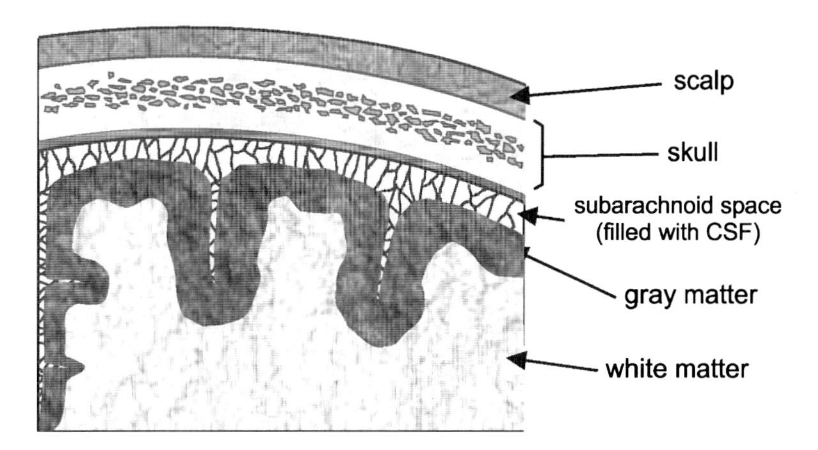

Fig. 3.3 Schematic diagram of the cross-section of the anatomical structure of the adult head: The head consists of various tissue components that have different optical properties. The thickness of the superficial tissues is not uniform, and the brain surface is folded with sulci.

Table 3.1 Typical optical properties of tissues of an adult head model at 800-nm wavelength

Tissue types (mm^{-1})	Transport scattering coefficient (mm^{-1})	Absorption coefficient (mm^{-1})
Scalp	1.9	0.018
Skull	1.6	0.016
Subarachnoid space (CSF)	0.24	0.004
Gray matter	2.2	0.036
White matter	9.1	0.014

Data chosen from the reported data for dermis [64], pig skull [62], CSF layer [16], and human brain [63]

The optical properties include the refractive index and the parameters describing the absorption and scattering characteristics of tissue. The absorption characteristic is described by the absorption coefficient, μ_a, which is related to the mean free path for absorption. The scattering characteristic is described by scattering coefficient μ_s, which is related to the mean free path for scattering, and the scattering phase function or anisotropy factor, g, which is the average of the cosine value of the scattering phase function. The phase function is the intensity distribution of scattered light as a function of the deflection angle.

A schematic diagram of the cross section of an adult head is shown in Fig. 3.3. Basically, the head has a layered structure – the scalp, skull, dura mater, arachnoid mater, the subarachnoid space filled with CSF, pia mater, gray matter and white matter – and they consist of variety types of tissues. There are several publications about the optical properties of biological tissue [61–66]. The optical properties of tissue samples can be measured in vitro using an integrating sphere system [62–64]. The optical properties of removed tissue samples are probably different from those measured in vivo. Although optical properties can be measured in vivo by time-, phase- and spatially resolved spectroscopies, it is difficult to independently measure a particular type of tissue [65]. Since the optical properties of limited types of tissue have been reported, the head model consists of five types of tissues, of which the transport scattering and absorption coefficients are shown in Table 3.1.

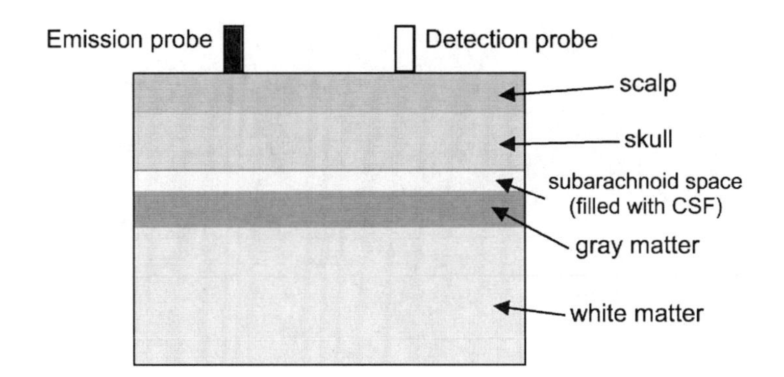

Fig. 3.4 **Simplified layered head model**: The simplified head model consists of five types of tissues. The tissue layers are parallel to each other

3.4.2 Simplified Head Model and Realistic Head Model

The simplified head model consists of the five types of tissues; however, the sophisticated geometry of the tissue structure is ignored and tissue layers are parallel to each other, as shown in Fig. 3.4. The simplified head model is adequate for evaluating the effect of superficial tissue thickness on light propagation in the brain [12, 16, 67].

In a real head, the thickness of superficial tissue, such as the scalp and skull, is not uniform and brain surface is folded with sulci. These sophisticated structures are important for precise analysis of photon migration in the brain [13, 17–21]. The anatomical geometry of a realistic head model is based upon MRI scans, as shown in Fig. 3.5a. The image is segmented into five types of tissue, as shown in Fig. 3.5b1–b5. The three-dimensional segmented geometry of the realistic head model is shown in Fig. 3.5c. The model consists of regular voxels, and each voxel is specified by its optical properties.

3.5 Simulation Methods of Photon Migration

3.5.1 Monte Carlo Method

There are several different mathematical techniques for describing photon migration in highly scattering tissue. Monte Carlo simulation is a stochastic method related to the physics of photon migration in scattering media. It has the advantages of being conceptually simple and of permitting direct handling of complex geometries and optical heterogeneity.

In the Monte Carlo method, random migrations of many photon packets are traced in order to predict light propagation in tissue. Migration of an individual photon packet is determined by the scattering coefficient, scattering phase function, absorption coefficient, refractive index, and random numbers. The successive scattering length (l_i) is determined by the scattering coefficient and a random number [8, 23]:

$$l_i = \frac{-\ln(\sigma)}{\mu_{s\,i}}, \tag{3.10}$$

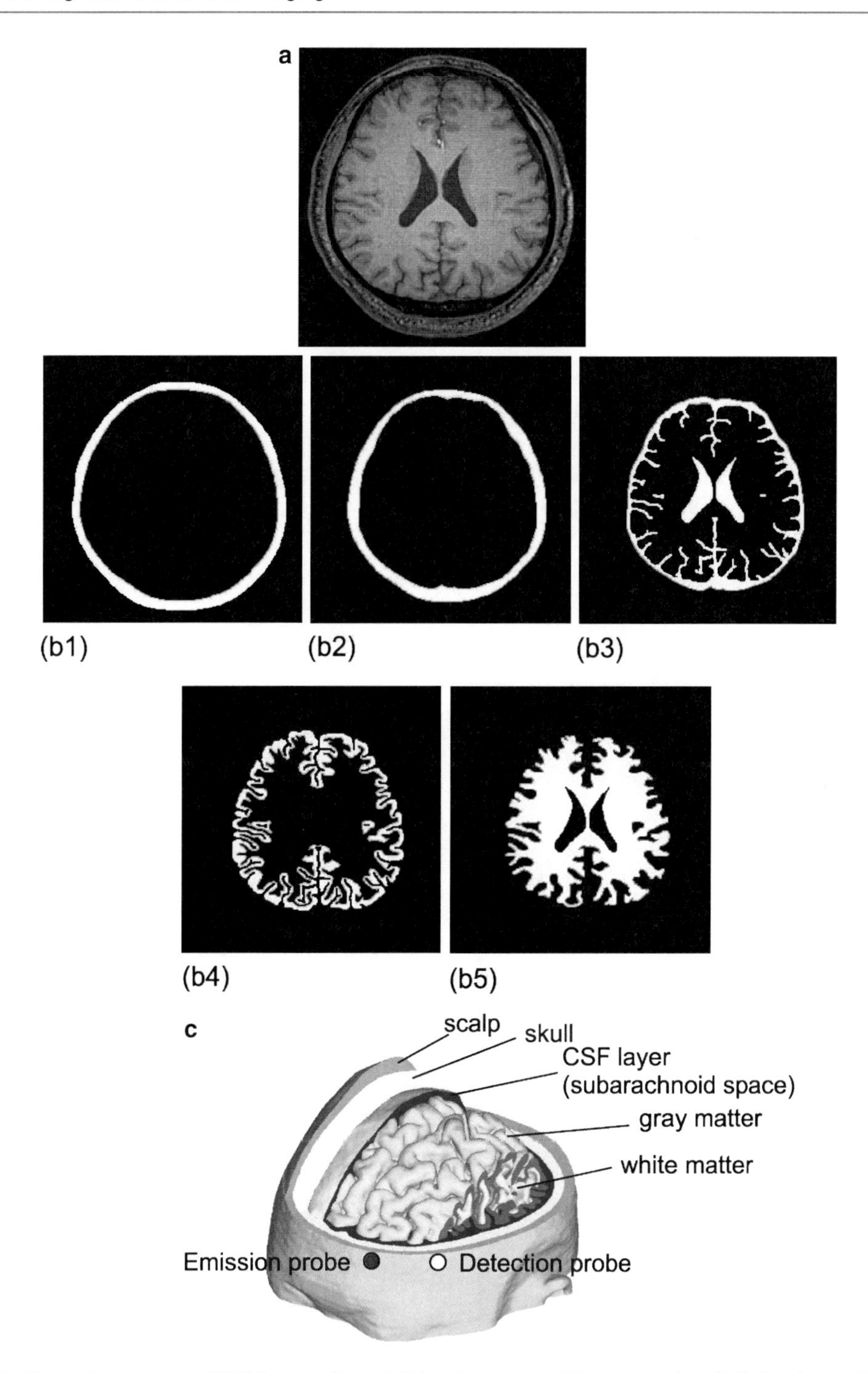

Fig. 3.5 Magnetic resonance (MR) image of an adult head, segmented images, and realistic head model: (**a**) MR image of an axial slice of an adult head. The MR image segmented into five types of tissues: (**b1**) scalp, (**b2**) skull, (**b3**) subarachnoid space (CSF layer), (**b4**) gray matter, and (**b5**) white matter. (**c**) The realistic head model consists of five types of tissues

where μ_{si} is the scattering coefficient of tissue type i and σ a random number between 0 and 1. The new direction of the photon packet is calculated from the scattering phase function and random numbers. The Henyey-Greenstein phase function given by Eq. 3.11 is frequently used to describe the probability distribution of the deflection angle of the photon packet (θ) in the Monte Carlo simulation [68]:

$$p(\cos\theta) = \frac{1 - g^2}{2(1 + g^2 - 2g\cos\theta)^{\frac{3}{2}}}, \tag{3.11}$$

where g is the anisotropy factor, which is defined as $<\cos\theta>$, the mean cosine of the deflection angle. The anisotropy factor is zero for isotropic scattering and unity for complete forward scattering. The anisotropic factor is typically in the range 0.7–0.95 for tissues in the near-infrared wavelength region. When a photon packet crosses the boundary from tissue type i to tissue type j, the scattering length in the second tissue (l_j) from the boundary is corrected by the scattering coefficients in each medium:

$$l_j = \frac{\mu_{s\,i}}{\mu_{s\,j}}(l_i - l'), \tag{3.12}$$

where $\mu_{s\,j}$ is the scattering coefficient of the second tissue j, and l' is the pathlength to the boundary from the previous scattering point. The influence of refractive index mismatch can be calculated from Fresnel's law and Snell's law.

The photon packet has a survival weight, which is related to light intensity. When the photon packet is scattered out of the head, the ultimate survival weight (W) is calculated from the absorption coefficient of each type of tissue ($\mu_{a\ i}$) and the accumulated pathlength (L_i) in each type of tissue L_i:

$$W = W_0 \exp\left(-\sum_{i=1}^{N} \mu_{a\,i} L_i\right), \tag{3.13}$$

where W_0 is the survival weight of the photon packet reduced only by reflection on the boundary due to refractive index mismatch.

There are other algorithms to determine successive scattering length and survival weight [22, 24]. Scattering length is determined by the sum of the scattering and absorption coefficients, and random number σ:

$$l_i = \frac{-\ln(\sigma)}{(\mu_{s\,i} + \mu_{a\,i})}. \tag{3.14}$$

The survival weight is recalculated at each scattering position:

$$W_n = \left(1 - \frac{\mu_{a\,i}}{\mu_{s\,i} + \mu_{a\,i}}\right) W_{n-1}, \tag{3.15}$$

where W_n and W_{n-1} are the survival weights at the present and previous scattering positions, respectively.

3.5.2 Diffusion Equation

A precise description of photon migration in tissue is provided by the radiative transport equation [45, 69]. Since numerical methods for solving the radiative transport equation are computationally expensive, a diffusion equation that approximates the transport equation has frequently been used to describe photon migration in a highly scattering tissue [45, 70]:

$$\frac{1}{c}\frac{\partial}{\partial t}\Phi(\mathbf{r},t) - \kappa(\mathbf{r})\nabla^2\Phi(\mathbf{r},t) + \mu_a\Phi(\mathbf{r},t) \quad = q_0(\mathbf{r},t), \tag{3.16}$$

where c is the speed of light, $q_0(\boldsymbol{r},t)$ is the distribution of the isotropic sources, $\Phi(\boldsymbol{r},t)$ is the photon density distribution, and κ is the diffusion coefficient, defined as [71]

$$\kappa(\mathbf{r}) = \frac{1}{3\mu'_s(\mathbf{r})}, \tag{3.17}$$

where μ'_s is the transport scattering coefficient:

$$\mu'_s(\mathbf{r}) = (1 - g(\mathbf{r}))\mu_s(\mathbf{r}), \tag{3.18}$$

where μ_s is the scattering coefficient and g is the anisotropy factor. The diffusion equation is valid under the appropriate conditions where $\mu'_s \gg \mu_a$ and $1/\mu'_s$ is much smaller than the distance between emission and detection points.

The diffusion equation can be analytically solved for only simple geometries such as a semiinfinite space, a sphere, and a layered slab [25, 27, 72]. Numerical methods are used to solve the diffusion equation, and the finite-element method is widely applied to analysis of photon migration in heterogeneous models with a sophisticated geometry [20, 28, 29, 42].

3.6 Influence of Superficial Tissues on Photon Migration in Brain

3.6.1 Influence of Optical Properties

The fundamental problem of NIRS imaging is the ambiguity in light propagation within the head caused by scattering of tissue. It is important to know the sensitivity of the NIRS signal to absorption change in the brain and the volume of tissue interrogated by the probe pairs. The heterogeneity of tissues in the head, especially the low-scattering CSF layer, has a strong effect on light propagation in the brain [12–16]. Photon migration in the three models is compared in order to evaluate the influence of optical heterogeneity in superficial tissues on light propagation in the brain. The heterogeneous structure of models (a) and (b) is generated from an MRI scan of the adult head, whereas model (c) has a homogeneous structure. Model (a) is a realistic head model whose optical properties are shown in Table 3.1. The boundary of each type of tissue in model (b) is the same as that in model (a); however, the optical properties of the CSF layer in model (b) is replaced by the optical properties of the skull. The transport scattering and absorption coefficients of homogeneous model (c) are 1.7 and 0.017 mm^{-1}, respectively. The optical properties of the homogeneous model are chosen such that mean optical pathlength for the homogeneous model at a probe spacing of 30 mm is the same as that for the realistic head model. Although there are no boundaries in model (c), hypothetical boundaries are drawn to compare penetration depth.

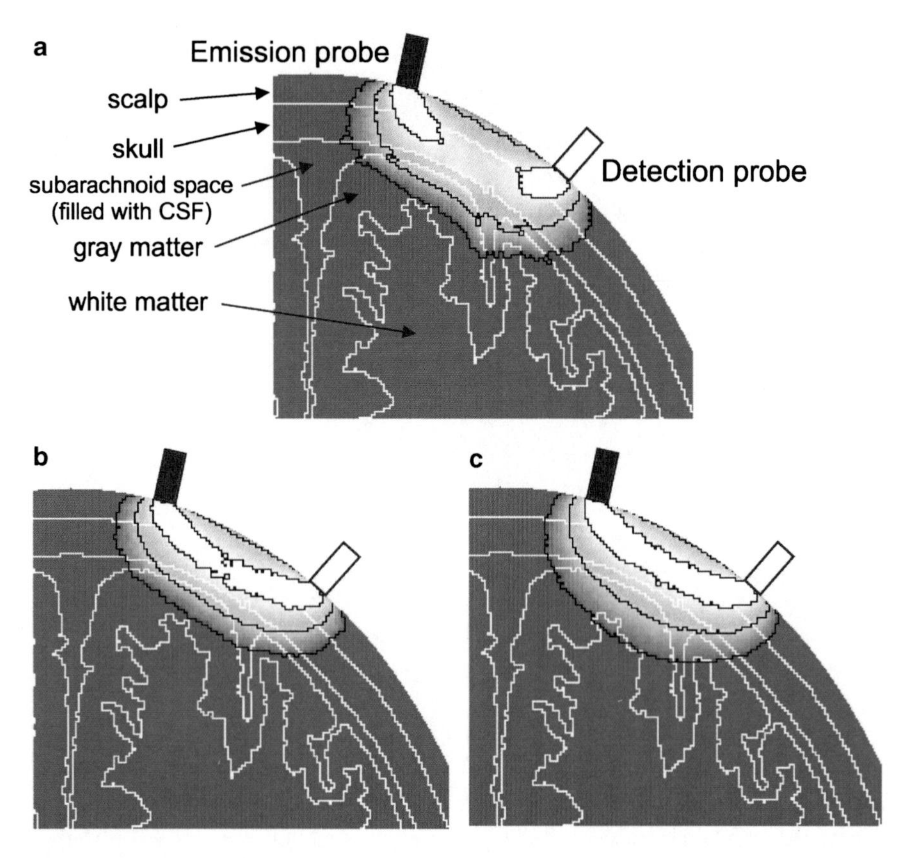

Fig. 3.6 Influence of the optical heterogeneity of superficial tissues on spatial sensitivity profile in realistic head models: Spatial sensitivity profile in transverse plane of the realistic head model for a probe spacing of 30 mm. Model (**a**) is a realistic head model whose optical properties are shown in Table 3.1. The boundary of each type of tissue in model (**b**) is the same as that in model (**a**); however, the optical properties of the CSF layer in model (**b**) are replaced by the optical properties of the skull. The transport scattering and absorption coefficients of homogeneous model (**c**) are 1.7 and 0.017 mm^{-1}, respectively. The colorscale of all the profiles is normalized by the maximum value in each result, and the contours are drawn for 50%, 10%, and 1% with respect to the maximum sensitivity point

The spatial sensitivity profiles in models (a), (b), and (c) for a probe spacing of 30 mm are depicted in Fig. 3.6. The spatial sensitivity profile in realistic head model (a) is obviously distorted around the CSF layer, whereas the spatial sensitivity profile in homogeneous model (c) forms a well-known banana shape. The spatial sensitivity profile in the realistic head model is broadly distributed in the shallow region of the brain. Although the difference between realistic head model (a) and model (b) is only the optical properties of the CSF layer, the spatial sensitivity profile in model (b) is similar to that in homogeneous model (c). These results indicate that the low-scattering CSF layer has an influence on the spatial sensitivity profile in the brain.

A comparison of mean optical pathlength for the head and partial optical pathlength in the brain as a function of probe spacing is shown in Fig. 3.7a, b, respectively. The mean optical pathlength, which can be measured by time- and phase-resolved experiments, is almost the same across the three models. The slope of the mean optical pathlength for the realistic head model slightly decreases beyond a probe spacing of 30 mm. The partial optical pathlength in the brain, which cannot be directly measured by experiments, for all the models starts to rise once probe spacing exceeds 20 mm. The slope of the partial optical pathlength for the realistic head model is obviously steeper than that of the other two models. The realistic head model with a low-scattering CSF layer has approximately

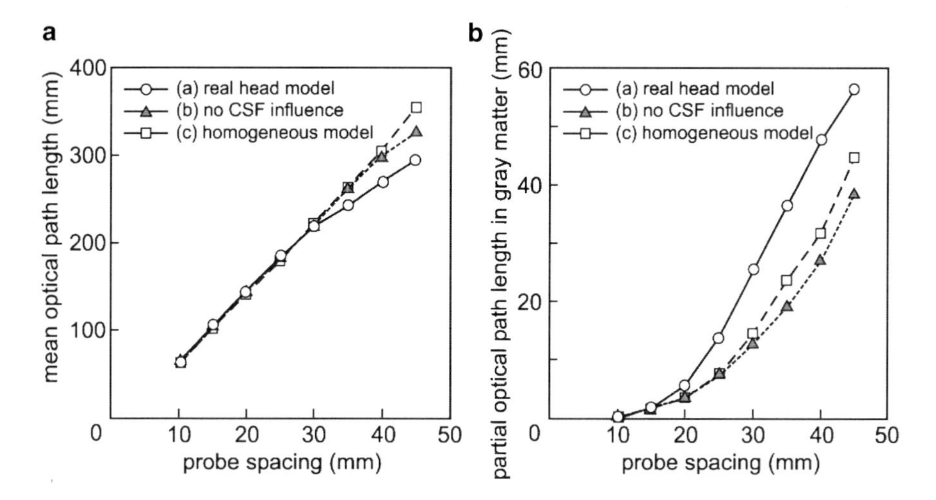

Fig. 3.7 Influence of the optical heterogeneity of superficial tissues on the mean optical pathlength in the whole head and partial optical pathlength in the brain: The optical properties of the models are the same as the models shown in Fig. 3.6. Partial optical pathlength in the brain is affected by the optical heterogeneity of superficial tissues

twice the partial optical pathlength in the gray matter compared to models without the low-scattering layer. This indicates that the sensitivity of the NIRS signal is improved by the presence of the low-scattering CSF layer. Optical heterogeneity, especially a low-scattering CSF layer, is important for modeling of photon migration for brain-function imaging by NIRS.

3.6.2 Influence of Thickness of Superficial Layers

The light that propagates into the brain must pass through superficial tissues, and hence the thickness of these superficial tissues affects the NIRS signal [67, 73]. The thickness of the skull significantly varies within the head and between individuals. Light propagation in the brain is strongly affected by the presence of a low-scattering CSF layer, and the thickness of the CSF layer is easily varied because the brain can move within the skull. Photon migration in the simplified head models that include the skull and CSF layers of various thicknesses is predicted by a Monte Carlo simulation in order to obtain partial optical pathlength in the gray matter and the spatial sensitivity profile. The optical properties of the layers in the model are shown in Table 3.1. The thicknesses of the scalp and gray matter are 3 and 4 mm, respectively. The thickness of the skull is changed from 4 to 12 mm, with the CSF thickness held constant at 2 mm, to evaluate the influence of skull thickness on light propagation in the brain. Similarly, the thickness of the CSF layer is changed from 0.25 to 5 mm with skull thickness held constant at 7 mm.

The influence of skull thickness on partial optical pathlength in gray matter is shown in Fig. 3.8: partial optical pathlength decreases with increasing skull thickness. In the case of a probe spacing of 30 mm, the model with a 7-mm-thick skull has almost three times the partial optical pathlength in the gray matter compared to the model with a 10-mm-thick skull. The sensitivity of the NIRS signal is significantly affected by skull thickness. Spatial sensitivity profiles for the head models with skulls of 7- and 10-mm thickness at a probe spacing of 30 mm are shown in Fig. 3.9a, b, respectively. Deeper penetration from the head surface is observed in the model with a thicker skull, although penetration into the brain slightly decreases with an increase in skull thickness, as shown in Fig. 3.9a1, b1.

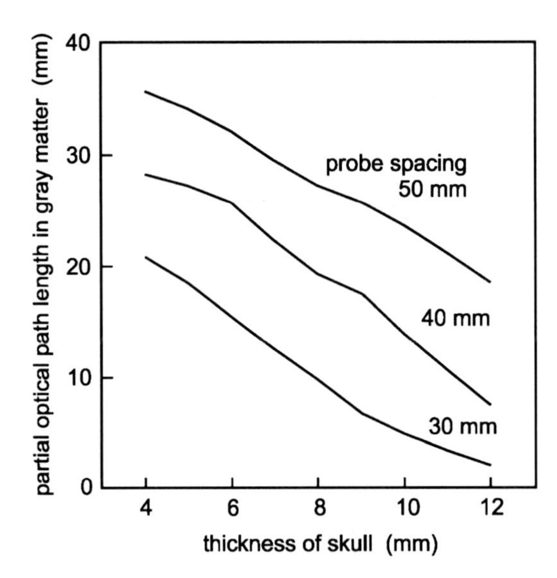

Fig. 3.8 Influence of skull thickness on partial optical pathlength in the brain: The thickness of the skull is changed from 4 to 12 mm, with the CSF thickness held constant at 2 mm. The partial optical pathlength in the brain decreases with increasing skull thickness

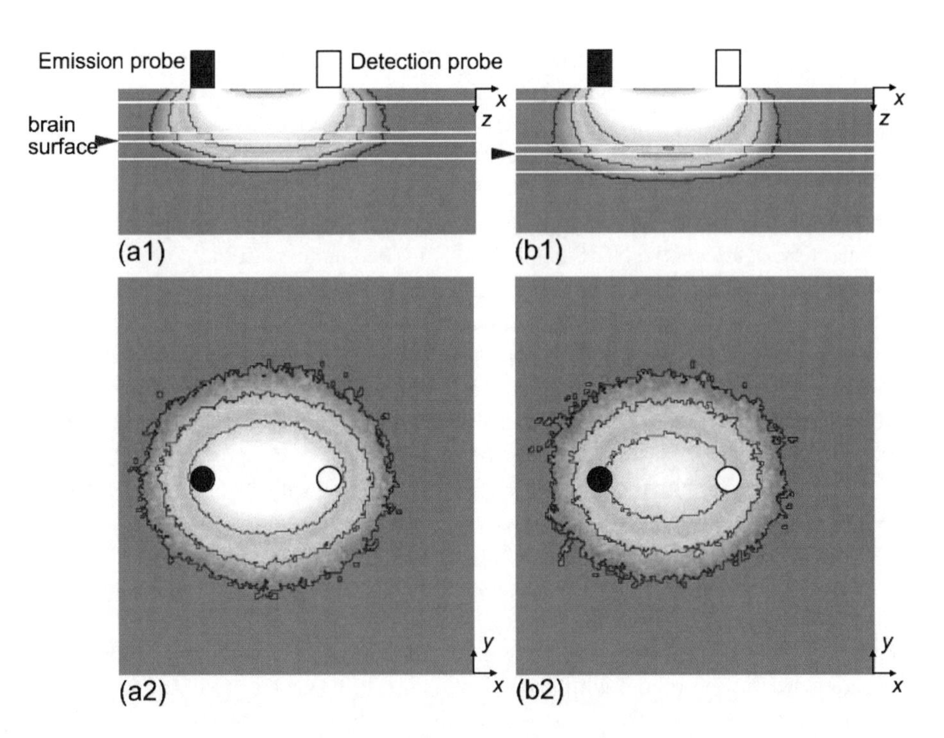

Fig. 3.9 Influence of skull thickness on spatial sensitivity profile: Spatial sensitivity profiles (**1**) in the x–z plane in the head model and (**2**) in the x–y plane in the gray matter at a probe spacing of 30 mm. The thicknesses of the skull are (**a**) 7 and (**b**) 10 mm. The colorscale is normalized by maximum value in (**a**). The contours are drawn for 10%, 1%, and 0.1% with respect to the maximum value

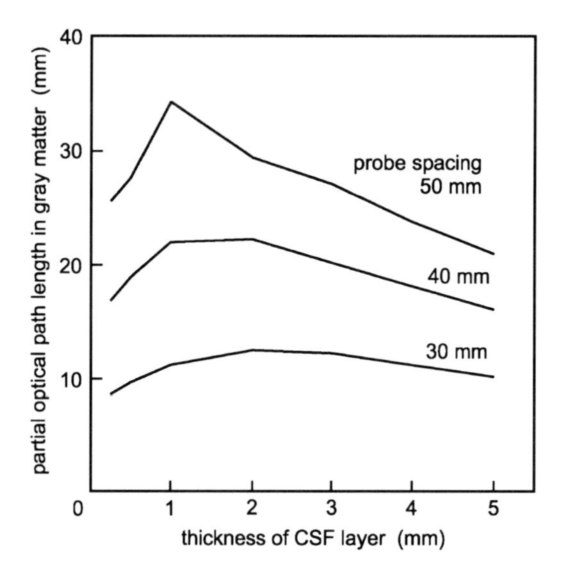

Fig. 3.10 Influence of CSF layer thickness on partial optical pathlength in the brain: The thickness of the CSF layer is altered from 0.25 to 5 mm, with the skull thickness held constant at 7 mm. The influence of CSF layer thickness on partial optical pathlength in the brain for a probe spacing of 30 mm is less significant than that of the skull

The most intense region of the spatial sensitivity profile in the *x–y* plane in gray matter is reduced and concentrated at the midpoint of the emission and detection probes with an increase in the skull thickness.

The influence of CSF layer thickness on partial optical pathlength in gray matter is shown in Fig. 3.10. The thickness of the CSF layer is altered from 0.25 to 5 mm with skull thickness held constant at 7 mm. Partial optical pathlength initially increases when the thickness of the CSF layer increases, and then starts to decrease with additional increases in CSF layer thickness. For a probe spacing of 30 mm, partial optical pathlength for the head model with a 2-mm-thick CSF layer is approximately 20% longer than that for a head model with a 5-mm-thick CSF layer. The difference in partial optical pathlength caused by the CSF layer thickness is less significant than that caused by skull thickness. However, CSF thickness can be varied by brain movement. Change in thickness of the CSF layer during measurement directly affects the NIRS signal. The spatial sensitivity profiles for the head models with CSF layers of 0.5- and 3-mm thicknesses at a probe spacing of 30 mm are shown in Fig. 3.11a, b, respectively. The penetration depth into the brain only slightly depends on CSF layer thickness. Both the maximum intensity region and the extreme contours of the spatial sensitivity profile in the *x–y* plane in the gray matter are broadened when CSF layer thickness increases. This tendency is not significant beyond a 3-mm CSF layer for a probe spacing of 30 mm.

3.6.3 Applicability of Diffusion Equation

The diffusion equation has an advantage in computation time compared to the Monte Carlo method and is widely used to simulate photon migration in tissue. The most serious drawback of the diffusion equation in its application to modeling of photon migration in the head is that diffusion approximation is not valid in a non- or low-scattering medium. The head has a subarachnoid space filled with non-scattering CSF around the brain. The CSF layer cannot be ignored when modeling photon migration in the head because the structure in which a low-scattering CSF layer is sandwiched by the highly scattering skull and gray matter strongly affects light propagation in the brain. For light scattered by

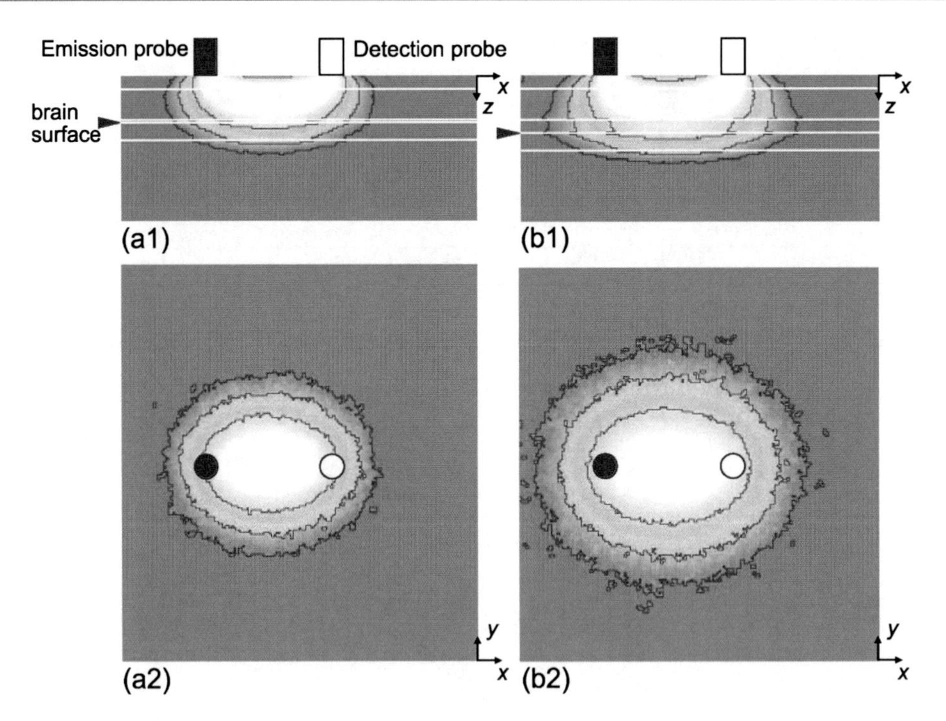

Fig. 3.11 **Influence of CSF layer thickness on spatial sensitivity profile**: Spatial sensitivity profiles (**1**) in the *x*–*z* plane in the head model and (**2**) in the *x*–*y* plane in the gray matter at a probe spacing of 30 mm. The thicknesses of the CSF layer are (**a**) 0.5 and (**b**) 3 mm. The colorscale is normalized by the maximum value in (**b**). The contours are drawn for 10%, 1%, and 0.1% with respect to the maximum value

weblike strands, arachnoid trabeculae, in the CSF layer, there are no experimental data on the optical properties of the CSF layer. The range of the scattering coefficient of the CSF layer used to calculate an adequate partial optical pathlength and spatial sensitivity profiles using the diffusion equation need to be evaluated [33, 34].

Photon migration in the simplified head model including a CSF layer of various scattering coefficients is now analyzed by the diffusion equation and is compared to that predicted by the Monte Carlo method for validating practical implementation of the diffusion equation. The thicknesses of the scalp, skull, CSF layer, and gray matter are 3, 7, 2, and 4 mm, respectively. The optical properties of each tissue (except for the transport scattering coefficient of the CSF layer) are shown in Table 3.1. The transport scattering coefficient of the CSF layer is changed from 0.01 to 1.0 mm^{-1}. Partial optical pathlengths in gray matter as a function of probe spacing calculated by the diffusion equation and the Monte Carlo method are shown in Fig. 3.12. The diffusion equation is solved by the finite-element method. The results calculated by the diffusion equation agree well with those predicted by the Monte Carlo method when the transport scattering coefficient of the CSF layer is 0.3 and 1.0 mm^{-1}. Partial optical pathlength in the gray matter calculated by the diffusion equation is obviously underestimated for the model with the CSF layer when the transport scattering coefficient is 0.01 mm^{-1}. The error in the detected intensity, mean optical pathlength, and partial optical pathlength in gray matter calculated by the diffusion theory for a probe spacing of 30 mm is shown in Fig. 3.13. The error in the results by the diffusion theory steeply increases with decreasing transport scattering coefficient when the transport scattering coefficient of the CSF layer is less than 0.2 mm^{-1}. It is difficult to accurately determine the transport scattering coefficient of the CSF layer because the density of the arachnoid traveculae may vary with position of the head. A practical and feasible assumption is that the transport scattering coefficient of the CSF layer is fixed at 0.3 mm^{-1} in order to adopt the diffusion equation for analysis of photon migration in the head models.

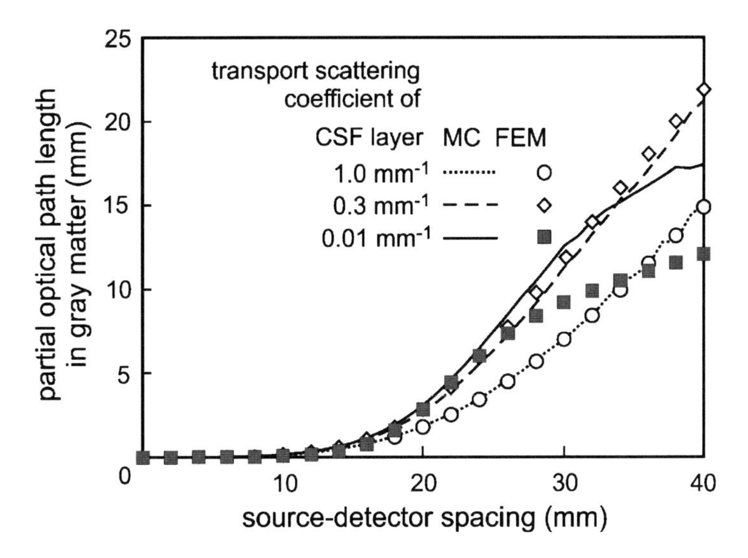

Fig. 3.12 Partial optical pathlength in the gray matter calculated by the diffusion equation and Monte Carlo method as a function of probe spacing: The diffusion equation is solved by the finite-element method. The results calculated by the diffusion equation agree with those obtained by the Monte Carlo method when the transport scattering coefficient of the CSF layer is greater than 0.3 mm^{-1}

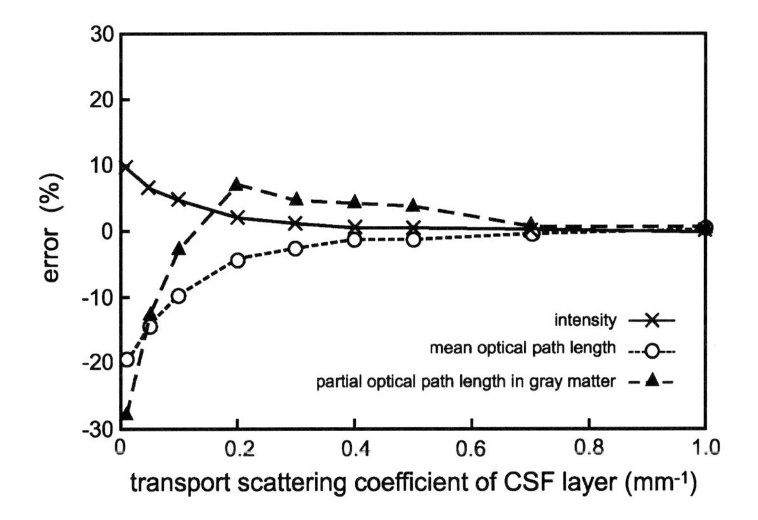

Fig. 3.13 Error in results calculated by the diffusion equation as a function of the transport scattering coefficient of the CSF layer for a probe spacing of 30 mm: The detected intensity, mean optical pathlength in the head, and partial optical pathlength in the gray matter calculated by the diffusion equation are compared to those by the Monte Carlo method. The diffusion equation is solved by the finite-element method

3.7 Modeling of NIRS Topography

A realistic head model can be used for evaluation of optimal probe arrangements and image reconstruction algorithms [20, 21]. The results of simulation of NIRS topography is shown in Fig. 3.14. The positions of the emission and detector probes and the activation region in the realistic head model are

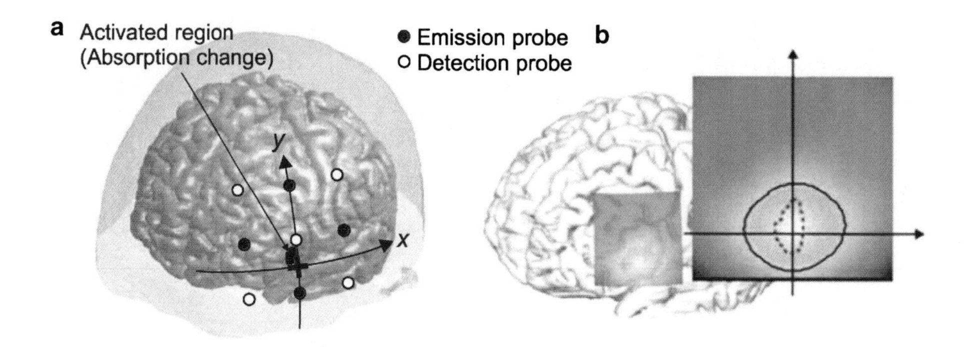

Fig. 3.14 Simulation of brain-function imaging by NIRS topography: (**a**) Photon migration in the realistic head model is predicted by the finite-element method in order to calculate the intensity change caused by absorption change. The four emission and five detector probes are alternatively attached at 30-mm interval lattice points on the scalp to construct 12 probe pairs. (**b**) The topographic image is calculated from the intensity changes by the mapping method. The *solid* and *broken lines* show the half-maximum of the measured topographic image and the actual absorption change in the model, respectively

shown in Fig. 3.14a. The four emission and five detector probes are alternatively attached at 30-mm interval lattice points on the scalp to construct 12 probe pairs. Photon migration in the head model is predicted by the finite-element method in order to calculate change in intensity caused by absorption change in the activated region of the brain detected by each probe pair. The topographic image is calculated from intensity changes by the mapping method, as shown in Fig. 3.14b. The result indicates that the size of the activated region in the topographic image is obviously greater than that of actual absorption change.

3.8 Summary

This chapter has presented photon migration in head models for brain-function imaging by NIRS. There are two basic types of NIRS imaging: tomography and topography. NIRS topography has been extensively applied to brain-function imaging because the image can be generated by the mapping method without solving the inverse problem. Modeling of photon migration in the head is important in order to solve the inverse problem to obtain the image by NIRS tomography and to improve the spatial resolution of NIRS topography. Light propagation in the brain is affected by the optical heterogeneity of superficial tissues, especially the low-scattering CSF layer. Spatial sensitivity profiles are distorted around the CSF layer. The sensitivity of the NIRS signal, which is indicated by partial optical pathlength in the brain, is improved by the presence of a low-scattering CSF layer. The sensitivity of the NIRS signal depends on the thickness of the superficial tissue. Since the thickness of the superficial tissues is not uniform and the brain surface is folded with sulci, the realistic head model in which the geometry is based on the anatomical structure of the head is important in order to accurately predict photon migration in the head. Photon migration in the head models can be calculated by the Monte Carlo method and approaches based on the diffusion equation. Although diffusion approximation is not valid in the low-scattering CSF layer, the results of photon migration in the head model calculated by the diffusion equation is practically feasible when the transport scattering coefficient of the CSF layer is 0.3 mm^{-1}.

Acknowledgments I would like to acknowledge funding support from the Japan Society for the Promotion of Science, Grant-in-Aid for Scientific Research (B) (19360035), and invaluable scientific discussions with Drs. Hiroshi Kawaguchi and Tsuyoshi Yamamoto.

Problems

3.1. Assume that the concentration of hemoglobin is changed from 0.1 to 0.11 mM and the oxygen saturation of the blood is changed from 65% to 70% in the activated region of the brain. The extinction coefficients of oxygenated hemoglobin and deoxygenated hemoglobin at 780-nm wavelength are 0.16 and 0.25 mM^{-1} mm^{-1}, respectively. The partial optical pathlength in the activated region for a probe pair is 5 mm.
   (a) Find the absorption change in the activated region.
   (b) Find the change in optical density (NIRS signal) caused by absorption change in the activated region.

3.2. Derive the equations that calculate the concentration change in oxygenated and deoxygenated hemoglobins from change in optical density (NIRS signal) at two wavelengths, λ_1 and λ_2. The extinction coefficient of oxygenated hemoglobin and deoxygenated hemoglobin is ε_{oxy-Hb} and $\varepsilon_{deoxy-Hb}$, respectively. Assume that the wavelength dependence of the partial optical pathlength in the activated region $<L_{act}>$ can be ignored.

3.3. Draw polar plots of the probability distribution of deflection angle $p(\theta)$ described by the Henyey-Greenstein phase function for $g = 0.1$, $g = 0.5$, and $g = 0.9$.

3.4. A pencil beam of short pulse is incident onto tissues, and diffusely reflected light is detected at 20 mm from the incident point. Analyze light propagation in the tissues by analytical solution of the diffusion equation described in [25]. The optical properties of the tissues: (1) $\mu_s = 10\ mm^{-1}$, $g = 0.9$, $\mu_a = 0.01\ mm^{-1}$. (2) $\mu_s = 10\ mm^{-1}$, $g = 0.85$, $\mu_a = 0.01\ mm^{-1}$. (3) $\mu_s = 5\ mm^{-1}$, $g = 0.8$, $\mu_a = 0.02\ mm^{-1}$. Although the diffusion coefficient is defined as $\kappa = 1/3\{(\mu'_s + \mu_a)\}$ in [25], $\kappa = 1/(3\mu'_s)$ can be used for the calculations. The speed of light in the medium is 0.2 mm/ps, and refractive index mismatch at the tissue boundary can be ignored.
   (a) Determine the transport scattering coefficient of each tissue.
   (b) Determine the depth of the isotropic point source created by the incident beam.
   (c) Draw the temporal distribution of reflectance.

Further Reading

Frostig RD (ed) (2002) In vivo optical imaging of brain function. CRC Press, New York/Washington, DC
Potter RF (ed) (1993) Medical optical tomography: functional imaging and monitoring. SPIE Press, Washington, DC
Tuchin V (2000) Tissue optics. SPIE Press, Washington, DC
Wang LV, Wu H (2007) Biomedical optics. Wiley, New York

References

1. Wyatt JS, Delpy DT, Cope M, Wray S, Reynolds EOR (1986) Quantification of cerebral oxygenation and haemodynamics in sick newborn infants by near infrared spectroscopy. Lancet 328(8515):1063–1066
2. Chance B, Leigh JS, Miyake H, Smith DS, Nioka S, Greenfeld R, Finander M, Kaufmann K, Levy W, Young M (1988) Comparison of time-resolved and -unresolved measurements of deoxyhemoglobin in brain. Proc Natl Acad Sci U S A 85(14):4971–4975
3. Hoshi Y, Tamura M (1993) Detection of dynamic changes in cerebral oxygenation coupled to neural function during mental work in man. Neurosci Lett 150(1):5–8
4. Obrig H, Wenzel R, Kohl W, Horst S, Wobst P, Steinbrink J, Thomas F, Villringer A (2000) Near-infrared spectroscopy: does it function in functional activation studies of the adult brain. Int J Psychophysiol 35(2–3):125–142

5. Fujiwara N, Sakatani K, Katayama Y, Murata Y, Hoshino T, Fukaya C, Yamamoto T (2004) Evoked-cerebral blood oxygenation changes in false-negative activation in BOLD contrast functional MRI of patients with brain tumors. Neuroimage 14(4):1464–1471
6. Gibson AP, Hebden JC, Arridge SR (2005) Recent advances in diffuse optical imaging. Phys Med Biol 50(4): R1–R43
7. Delpy DT, Cope M, van der Zee P, Arridge SR, Wray S, Watt JS (1988) Estimation of optical pathlength through tissue from direct time of flight measurement. Phys Med Biol 33(12):1433–1442
8. Hiraoka M, Firbank M, Essenpreis E, Cope M, Arridge SR, van der Zee P, Delpy DT (1993) A Monte Carlo investigation of optical pathlength in inhomogeneous tissue and its application to near-infrared spectroscopy. Phys Med Biol 38(12):1859–1876
9. Okada E, Firbank M, Delpy DT (1995) The effect of overlying tissue on the spatial sensitivity profile of near-infrared spectroscopy. Phys Med Biol 40(12):2093–2108
10. Okada E, Schweiger M, Arridge SR, Firbank M, Delpy DT (1996) Experimental validation of Monte Carlo and finite-element methods for the estimation of the optical pathlength in inhomogeneous tissue. Appl Opt 35(19):3362–3371
11. Hielscher AH, Liu H, Chance B, Tittel FK, Jacques SL (1996) Time-resolved photon emission from layered turbid media. Appl Opt 35(4):719–728
12. Okada E, Firbank M, Schweiger M, Arridge SR, Cope M, Delpy DT (1997) Theoretical and experimental investigation of near-infrared light propagation in a model of the adult head. Appl Opt 36(1):21–31
13. Firbank M, Okada E, Delpy DT (1998) A theoretical study of the signal contribution of regions of the adult head to near-infrared spectroscopy studies of visual evoked responses. Neuroimage 8(1):69–78
14. Wolf M, Keel M, Dietz V, von Siebenthal K, Bucher HU, Baenziger O (1999) The influence of a clear layer on near-infrared spectrophotometry measurements using a liquid neonatal head phantom. Phys Med Biol 44(7):1743–1754
15. Okada E (2000) The effect of superficial tissue of the head on spatial sensitivity profiles for near-infrared spectroscopy and imaging. Opt Rev 7(5):375–382
16. Okada E, Delpy DT (2003) Near-infrared light propagation in an adult head model, I: modeling of low-level scattering in the cerebrospinal fluid layer. Appl Opt 42(16):2906–2914
17. Boas DA, Culver JP, Stott JJ, Dunn AK (2002) Three-dimensional Monte Carlo code for photon migration through complex heterogeneous media including the adult human head. Opt Express 10(3):159–170
18. Fukui Y, Ajichi Y, Okada E (2003) Monte Carlo prediction of near-infrared light propagation in realistic adult and neonatal head models. Appl Opt 42(16):2881–2887
19. Fukui Y, Yamamoto T, Kato T, Okada E (2003) Analysis of light propagation in a three-dimensional realistic head model for topographic imaging by finite-difference method. Opt Rev 10(5):470–473
20. Kawaguchi H, Koyama T, Okada E (2007) Effect of probe arrangement on reproducibility of images by near-infrared topography evaluated by a virtual head phantom. Appl Opt 46(10):1658–1668
21. Heiskala J, Hiltunen P, Nissilä I (2009) Significance of background optical properties, time-resolved information and optode arrangement in diffuse optical imaging of term neonates. Phys Med Biol 54(3):535–554
22. Wilson BC, Adam G (1983) A Monte Carlo model for the absorption and flux distributions of light in tissue. Med Phys 10(6):824–830
23. van der Zee P, Delpy DT (1987) Simulation of the point spread function for light in tissue by a Monte Carlo technique. Adv Exp Med Biol 215:179–191
24. Wang L, Jacques SL, Zheng L (1995) MCML – Monte Carlo modeling of light transport in multi-layered tissues. Comput Methods Programs Biomed 47(2):131–146
25. Patterson MS, Chance B, Wilson BC (1989) Time-resolved reflectance and transmittance for the noninvasive measurement of tissue optical properties. Appl Opt 28(12):2331–2336
26. Farrell TJ, Patterson MS, Wilson B (1992) A diffusion theory model of spatially resolved, steady-state diffuse reflectance for the noninvasive determination of tissue optical properties in vivo. Med Phys 19(4):879–888
27. Arridge SR, Cope M, Delpy DT (1992) Theoretical basis for the determination of optical pathlengths in tissue: temporal and frequency analysis. Phys Med Biol 37(7):1531–1560
28. Arridge SR, Schweiger M, Hiraoka M, Delpy DT (1993) A finite-element approach for modeling photon transport in tissue. Med Phys 20(2):299–309
29. Yamada Y, Hasegawa Y (1993) Time-resolved FEM analysis of photon migration in random media. Proc SPIE 1888:167–178
30. Firbank M, Arridge SR, Schweiger M, Delpy DT (1996) An investigation of light transport through scattering bodies with non-scattering regions. Phys Med Biol 41(4):767–783
31. Dehghani H, Arridge SR, Schweiger M, Delpy DT (2000) Optical tomography in the presence of void regions. J Opt Soc Am A 17(9):1659–1670
32. Hayashi T, Kashio Y, Okada E (2003) Hybrid Monte Carlo diffusion method for light propagation in tissue with a low-scattering region. Appl Opt 42(16):2888–2896

33. Koyama T, Iwasaki A, Ogoshi Y, Okada E (2005) Practical and adequate approach to modeling light propagation in an adult head with low-scattering regions by use of diffusion theory. Appl Opt 44(11):2094–2103
34. Custo A, Wells WM III, Barnett AH, Hillman EMC, Boas DA (2006) Effective scattering coefficient of the cerebral spinal fluid in adult head models for diffuse optical imaging. Appl Opt 45(19):4747–4755
35. Oki Y, Kawaguchi H, Okada E (2009) Validation of practical diffusion approximation for virtual near infrared spectroscopy using a digital head phantom. Opt Rev 16(2):153–159
36. Schotland JC, Haselgrove JC, Leigh JS (1993) Photon hitting density. Appl Opt 32(4):448–453
37. Arridge SR (1995) Photon-measurement density functions, 1: analytical forms. Appl Opt 34(31):7395–7409
38. Eda H, Oda I, Ito Y, Wada Y, Oikawa Y, Tsunazawa Y, Takada M (1999) Multichannel time-resolved optical tomographic imaging system. Rev Sci Instrum 70(9):3595–3602
39. Schmidt FEW, Fry ME, Hillman EMC, Hebden JC, Delpy DT (2000) A 32-channel time-resolved instrument for medical optical tomography. Rev Sci Instrum 71(1):256–265
40. Wabnitz H, Moeller M, Liebert A, Obrig H, Steinbrink J, Macdonald R (2010) Time-resolved near-infrared spectroscopy and imaging of the adult human brain. Adv Exp Med Biol 662:143–148
41. Arridge SR (1993) The forward and inverse problems in time-resolved infrared imaging. In: Müller G et al (eds) Medical optical tomography: functional imaging and monitoring. SPIE Press, Bellingham, pp 35–64
42. Schweiger M, Arridge SR, Delpy DT (1993) Application of the finite-element method for the forward and inverse models in optical tomography. J Math Imaging Vis 3(3):263–283
43. Arridge SR, Schweiger M (1995) Photon-measurement density functions, 2: finite-element-method calculations. Appl Opt 34(34):8026–8037
44. Arridge SR, Hebden JC (1997) Optical imaging in medicine, II: modelling and reconstruction. Phys Med Biol 42(5):841–853
45. Arridge SR (1999) Optical tomography in medical imaging. Inverse Probl 15(2):R41–R93
46. Gibson AP, Hebden JC, Riley J, Everdell N, Schweiger M, Arridge SR, Delpy DT (2005) Linear and nonlinear reconstruction for optical tomography of phantoms with nonscattering regions. Appl Opt 44(19):3925–3936
47. Gibson AP, Austin T, Everdell NL, Schweiger M, Arridge SR, Meek JH, Wyatt JS, Delpy DT, Hebden JC (2006) Three-dimensional whole-head optical tomography of passive motor evoked responses in the neonate. Neuroimage 30(2):521–528
48. Austin T, Gibson AP, Branco G, Yusof R, Arridge SR, Meek JH, Wyatt JS, Delpy DT, Hebden JC (2006) Three-dimensional optical imaging of blood volume and oxygenation in the preterm brain. Neuroimage 31(4):1426–1433
49. Maki A, Yamashita Y, Ito Y, Watanabe E, Mayanagi Y, Koizumi H (1995) Spatial and temporal analysis of human motor activity using noninvasive NIR topography. Med Phys 22(12):1997–2005
50. Koizumi H, Yamamoto T, Maki A, Yamashita Y, Sato H, Kawaguchi H, Ichikawa N (2003) Optical topography: practical problems and new applications. Appl Opt 42(16):3054–3062
51. Watanabe E, Maki A, Kawaguchi F, Takashiro K, Yamashita Y, Koizumi H, Mayanagi Y (1998) Noninvasive assessment of language dominance with near-infrared spectroscopic mapping. Neurosci Lett 256(1):49–52
52. Taga G, Konishi Y, Maki A, Tachibana T, Fujiwara M, Koizumi H (2000) Spontaneous oscillation of oxy- and deoxy-hemoglobin changes with a phase difference throughout the occipital cortex of newborn infants observed using noninvasive optical topography. Neurosci Lett 282(1–2):101–104
53. Miyai I, Tanabe HC, Sase I, Eda H, Oda I, Konishi I, Tsunazawa Y, Suzuki T, Yanagida T, Kubota K (2001) Cortical mapping of gait in humans: a near-infrared spectroscopic topography study. Neuroimage 14(5):1186–1192
54. Suto T, Fukuda M, Ito M, Uehara T, Mikuni M (2004) Multichannel near-infrared spectroscopy in depression and schizophrenia: cognitive brain activation study. Biol Psychiatry 55(5):501–511
55. Yamamoto T, Maki A, Kadoya A, Tanikawa Y, Yamada Y, Okada E, Koizumi H (2002) Arranging optical fibres for the spatial resolution improvement of topographic images. Phys Med Biol 47(18):3429–3440
56. Kawaguchi H, Hayashi T, Kato T, Okada E (2004) Theoretical evaluation of accuracy in position and size of brain activity obtained by near-infrared topography. Phys Med Biol 49(12):2753–2765
57. Tian F, Alexandrakis G, Liu H (2009) Optimization of probe geometry for diffuse optical brain imaging based on measurement density and distribution. Appl Opt 48(13):2496–2504
58. Boas DA, Dale AM, Franceschini MA (2004) Diffuse optical imaging of brain activation: approaches to optimizing image sensitivity, resolution, and accuracy. Neuroimage 23(S1):S275–S288
59. Schweiger M, Arridge SR (1999) Optical tomographic reconstruction in a complex head model using a priori region boundary information. Phys Med Biol 44(11):2703–2721
60. Hielscher AH, Bluestone AY, Abdoulaev GS, Klose AD, Lasker J, Stewart M, Netz U, Beuthan J (2002) Near-infrared diffuse optical tomography. Dis Markers 18(5–6):313–337
61. Cheong WF, Prahl SA, Welch AJ (1990) A review of the optical properties of biological tissues. IEEE J Quantum Electron 26(12):2166–2185
62. Firbank M, Hiraoka M, Essenpreis M, Delpy DT (1993) Measurement of the optical properties of the skull in the wavelength range 650–950 nm. Phys Med Biol 38(4):503–510

63. van der Zee P, Essenpreis M, Delpy DT (1993) Optical properties of brain tissue. Proc SPIE 1888:454–465
64. Simpson CR, Kohl M, Essenpreis M, Cope M (1998) Near-infrared optical properties of ex vivo human skin subcutaneous tissue measured using the Monte Carlo inversion technique. Phys Med Biol 43(9):2465–2478
65. Kienle A, Glanzmann T (1999) In vivo determination of the optical properties of muscle with time-resolved reflectance using a layered model. Phys Med Biol 44(11):2689–2702
66. Meinke M, Müller G, Helfmann J, Friebel M (2007) Empirical model functions to calculate hematocrit-dependent optical properties of human blood. Appl Opt 46(10):1742–1753
67. Okada E, Delpy DT (2003) Near-infrared light propagation in an adult head model, II: effect of superficial tissue thickness on the sensitivity of the near-infrared spectroscopy signal. Appl Opt 42(16):2915–2922
68. Henyey LG, Greenstein JL (1941) Diffuse radiation in the galaxy. Astrophys J 93(1):70–83
69. Patterson MS, Wilson BC, Wyman DR (1991) The propagation of optical radiation in tissue, I: models of radiation transport and their application. Lasers Med Sci 6(2):155–168
70. Ishimaru A (1978) Wave propagation and scattering in random media. Academic, New York
71. Furutsu K, Yamada Y (1994) Diffusion approximation for a dissipative random medium and the applications. Phys Rev E 50(5):3634–3640
72. Martelli F, Sassaroli A, Yamada Y, Zaccanti G (2002) Analytical approximate solutions of the time-domain diffusion equation in layered slab. J Opt Soc Am A 19(1):71–80
73. Wang S, Shibahara N, Kuramashi D, Okawa S, Kakuta N, Okada E, Maki A, Yamada Y (2010) Effect of spatial variation of skull and cerebrospinal fluid layers on optical mapping of brain activity. Opt Rev 17(4):410–420

4 Clinical Application of NIRS

Miura Hajime

4.1 Introduction

The possibility of a diagnostic method based on near-infrared light has been discussed for more than 80 years. Since Jöbsis's introduction, near-infrared spectroscopy (NIRS) has been increasingly applied to measure tissue oxygenation in humans by detecting the levels of hemoglobin (Hb) and myoglobin (Mb) [1]. In this chapter, the recent advances in and prospects for clinical applications of NIRS are summarized.

There are several significant advantages that NIRS might offer over existing radiological techniques. The most obvious feature is its non-ionizing nature, which humans can be exposed to for a long time without harm due to accumulation of x-rays [2]. Similarly, NIRS does not lead to excess heat, unlike ultrasound, for which heating of tissues presents a significant hazard [2]. NIRS methods have potential uses as diagnostic tools because they can offer the ability to differentiate between human tissues with varying optical absorption or scatter. They can also allow functional information to be obtained based on specific absorption pattern. Therefore, NIRS can measure oxygenation in human tissues without the use of radioisotopes or other contrast agents. Nevertheless, NIRS has some disadvantages. One problem is a lack of quantification. There are options to quantify absolute changes in Hb levels, but it is not possible to measure these changes with NIRS. An additional magnetic resonance imaging (MRI) or computed tomography (CT) scan is necessary to quantify such changes. Another problem is difficulty in using complex formulas to convert measured optical density in order to quantify the amount of Hb circulating in the blood. These formulas have some limitations: due to their complexity, a small amount of interference can lead to data inaccuracies [3].

In order to appreciate the obstacles associated with the development of the NIRS method, the nature of the interactions that occur between near infrared light and human tissues should be considered. The most dominant interaction at near infrared wavelengths is elastic scattering. The scatter coefficient, μ_s, of many tissues has been measured at a variety of near-infrared wavelengths. Measurements of transmitted intensity are greatly affected by scattered light. Choice of wavelength is complicated by a need to consider the relative optical features of different tissues, the availability of suitable sources, and detector sensitivity detectors [1, 2]. The characteristic scatter of tissue is expressed in terms of transport scatter coefficient μ'_s, and the absorption coefficient is expressed by μ_a.

M. Hajime, Ph.D. (✉)
Laboratory for Applied Physiology, University of Tokushima, Tokushima 770-8502, Japan
e-mail: hajime-m@ias.tokushima-u.ac.jp

T. Jue and K. Masuda (eds.), *Application of Near Infrared Spectroscopy in Biomedicine*, Handbook of Modern Biophysics 4, DOI 10.1007/978-1-4614-6252-1_4,

Early applications of NIRS to measure oxygenation in various human tissues were made by Jöbsis [1] and Chance et al. [4]. The NIRS methodology involves detection of differences in light absorption from tissues and is comprised of different approaches, including: (1) continuous-wavelength spectroscopy [5], (2) spatially resolved spectroscopy [6], (3) time-resolved spectroscopy [7], and (4) phase-modulation spectroscopy [8]. The most versatile and widely used approach is continuous-wavelength spectroscopy. Near-infrared light in the range of 700–900 nm is used in an NIRS device because of its ability to penetrate into tissue.

To obtain a reflected-light signal from human tissues, the light source and photodetector (PD) are typically separated by 2.5–3.0 cm. The near-infrared light penetrates in a shallow arc, and the penetration depth is half the separation between the light source and PD (approximately 1.25–1.50 cm) [4, 6]. This indicates that longer separation distances result in deeper penetrations, but also translate into less light reaching the detector. The actual penetration depth and wavelength are very difficult to measure and vary with Hb concentration and tissue thickness (e.g., skin, fat, skull) between the light source and PD, as well as by human tissue type [9, 10]. This is the primary difference between the use of NIRS with reflected light and with transmitted light (such as pulse oximeters, where the light pathlength is known).

NIRS has been applied to measure oxygenation in a variety of human tissues, including muscle, blood vessels, brain, and connective tissue, and more recently has been used in a clinical setting to assess circulatory and metabolic abnormalities [1, 2].

4.2 Muscle Metabolism

Pulmonary ventilation supplies the alveoli with fresh air, which contains a high concentration of oxygen. The oxygen, combined with Hb, is transported to tissues via the circulation, and then consumed within tissues (e.g., muscle). A noninvasive method for oxidative phosphorylation is required to evaluate this process in humans. Nuclear magnetic resonance spectroscopy (NMRS) is currently used to measure the concentration of phosphorus atoms to monitor phosphocreatine, phosphate, ATP, and hydrogen ion concentrations [11, 12].

NIRS, however, offers another technique for indirectly measuring human muscle metabolism [11, 12]. It monitors tissue oxygen level by measuring optical absorption changes in oxygenated and deoxygenated Hb/Mb, and it allows for noninvasive measurement of the balance between oxygen consumption and oxygen supply. NIRS is assumed to map vascular PO_2, and NMR monitors deoxygenated Mb as a reflection of intracellular PO_2. A combined NIRS and NMR study has demonstrated that deoxygenated Mb kinetics from NMR matched those observed with NIRS from gastrocnemius muscle under planter flexion or pressure cuffing [13]. These results establish the feasibility and methodology to observe deoxygenated Mb and Hb signals in skeletal muscle and help to clarify the origin of the NIRS signal in skeletal muscle.

The oxygen uptake estimated from pulmonary gas exchange parameters has been used as an indirect index of muscle metabolism. NIRS has made it possible to directly and noninvasively measure localized O_2 extraction/consumption. A number of studies have used NIRS to measure the muscle oxygenation (Hb saturation and blood volume) of working/exercising muscles [14–21]. Figure 4.1 shows typical changes in oxygenated Hb that occur during cycling at five constant work rates [14]. With increasing work rate, oxygenated Hb at the vastus lateralis muscle decreased progressively. At the beginning of exercise, oxygenated Hb increased due to increased cardiac output and muscle blood flow. Thereafter, with increasing exercise intensity, it decreased due to increased oxygen consumption and oxygen extraction by working muscle. During recovery, oxygenated Hb increased rapidly due to

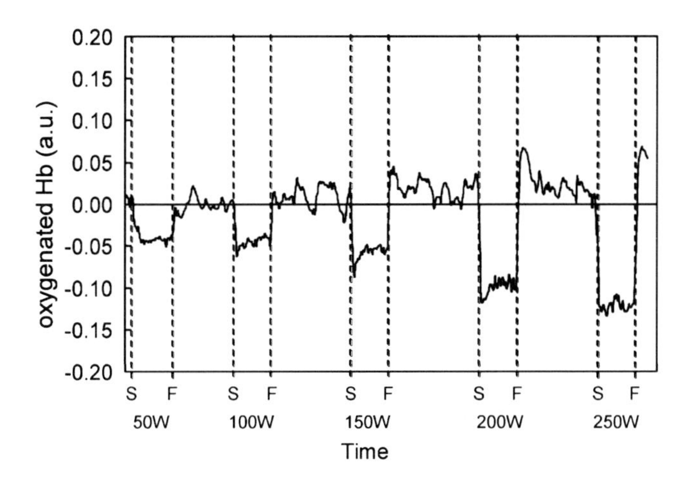

Fig. 4.1 **Continuous recording of the oxygenated Hb trace of vastus lateralis muscle during incremental cycling**: The symbol "*S*" indicates the starting point for each exercise and "*F*" the finishing point for each exercise. Note that a progressively decreasing signal (Hb desaturation) in the vastus lateralis muscle was found with increasing exercise intensity [14]

greater oxygen supply to muscle (hyperemia). Previous studies have measured oxygenation in the thigh or calf muscles during cycling [14], rowing [15], skating [16], knee extension [17], ankle plantar flexion [9, 18], and in arm muscles during arm cranking [19], elbow flexion [20], and finger flexion/hand grip exercise [21].

NIRS measures muscle oxygenation in the small blood vessels, or capillaries [11, 12]. Muscle oxygenation has been shown to correlate with localized muscle activity, as well as myoelectric activity and changes in blood lactate concentration [14]. NIRS measurements have also been correlated with phosphocreatine measurements taken using NMRS [22].

Some studies, however, have reported heterogeneity for muscle metabolism determined by multiple NIRS imaging devices [23–25]. As shown in Fig. 4.2, within the medial head of the gastrocnemius muscle during standing planter flexion, oxygenation was not uniformly distributed throughout the muscle, but instead was consistently different in the proximal and distal areas; the distal portion had greater deoxygenation and a higher decrease in blood volume during exercise, and it had greater oxygenation and increase in blood volume compared to the proximal portion [25].

4.2.1 Exercise Prescription/Training Effects

In order to keep/improve physical fitness or improve athletic performance, these NIRS measurements about muscle metabolism could be applied to prescribe exercise programs. Exercise intensity is usually determined by heart rate, oxygen uptake, or blood lactate concentration. These indicators can be used as a gauge of relative whole body exercise intensity. Changes in integrated myoelectric (iEMG) activity and degree of Hb saturation by NIRS in the vastus lateralis muscle were measured simultaneously during cycling with different constant work rates [14]. Changes in oxygenated Hb were observed to significantly correlate to iEMG for each subject ($r = -0.947$ to -0.993). The iEMG value has been used as an index for measuring the amount of muscle activity, degree of physical work, or amount of fatigue [26]. Therefore, it follows that Hb saturation levels provide reliable information regarding localized muscle activity during various exercise modes where many muscles are involved, such as arm and leg exercises.

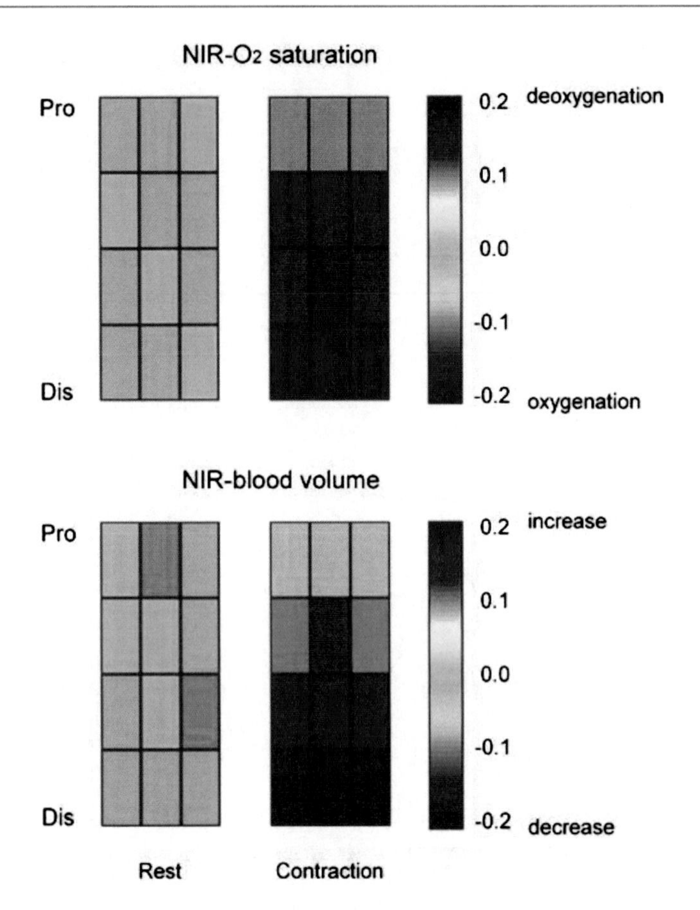

Fig. 4.2 Changes in oxygen saturation and blood volume during rest and muscle contraction: The symbol "*pro*" indicates the proximal portion and "*dis*" the distal portion of the medial head of the gastrocnemius muscle. Note that compared with the proximal portion, lower oxygen saturation and blood volume (sum of oxygenated Hb and deoxygenated Hb) were found during muscle contraction in the medial head of the gastrocnemius muscle [25]

Muscle oxygenation is also useful for evaluating either training effects or fitness level. The time course of post-exercise muscle oxygenation has been reported to be related to aerobic function [27, 28] and is accelerated after training in relation to changes in muscle oxidative enzymes or blood flow/ capillarization [29]. Puente-Maestu et al. [27] reported the speed of the re-oxygenation determined by NIRS during recovery from exercise over vastus lateralis muscle to be positively modified by endurance training, thus indicating it to be related to changes in oxidative enzyme (citrate synthase). However, muscle oxygenation has also been monitored during functional electrical stimulation in patients with spinal cord injury [30] and in healthy subjects for improving muscle strength [31].

In summary, these NIRS approaches might be useful for evaluating localized exercise intensity and can be applied in determining an optimal exercise program for athletes or for rehabilitation of patients with paraplegia.

4.2.2 Congestive Heart Failure and Q_{10} Deficiency

The clinical uses of NIRS muscle findings in varied diseases have also been explored. Booushel et al. [32] applied NIRS in clinical evaluation of patients with idiopathic muscle myalgia and exercise intolerance to ascertain the extent to which this clinical condition is associated with altered oxidative metabolism. Other investigators have shown that patients with congestive heart failure desaturate their muscle at a lower work level compared to healthy subjects, indicating an insufficient blood flow to exercising muscle [33, 34].

Patients with coenzyme Q_{10} deficiency related to limited muscle oxidative phosphorylation also have an oxygenation pattern during exercise different from that in healthy subjects [34]. In fact, the impairment in muscle oxidative phosphorylation is so severe that there is no increase in arterial-to-venous oxygen difference [35]. Accessory respiratory muscle oxygenation during exercise has also been assessed in patients with heart failure and in heart transplant recipients [36, 37].

Compartment syndrome of the lower limb is a clinical phenomenon that is characterized by an increase in compartment pressure in the lower, leading to circulatory deterioration and inadequate tissue oxygenation [38]. Arató et al. [39] reported that parallel observation of a subject's clinical status, intracompartmental pressure, and tissue oxygen saturation in the gastrocnemius muscle as determined by NIRS might lead to a correct and reliable diagnosis, thus resulting in optimal timing of fasciotomy and proper evaluation of the efficiency of surgery. Measurements of both the intracompartmental pressure and tissue oxygen saturation can be helpful in making an accurate diagnosis, so that an optimal determination can be made about whether or not a fasciometomy is indicated, and in evaluating a patient's status during postoperative follow-up.

4.3 Vessel Function

Several prospective studies have focused on the ability of arterial function to predict cardiovascular risk, which would allow for detection of atherosclerosis before the occurrence of critical events [40, 41]. Although they comprise the standard for a diagnosis of hypertension, systolic/diastolic blood pressure measurements are unsuitable for evaluating arterial function in the early stages of cardiovascular disease. Arterial compliance is determined by the functional and structural components related to the intrinsic elastic properties of arteries [42]. In order to evaluate endothelial function, it is necessary to measure the arterial structure, which can be carried out via cold pressor or flow-mediated dilation tests. These methods involve complicated techniques and take a long time.

The ankle-to-brachial index (ABI), however, is the standard for a diagnosis of arteriosclerosis obliterans (ASO). The ABI is determined using the systolic and brachial systolic blood pressure in the ankle. In order to conclusively evaluate/diagnose arteriosclerosis obliterans, it is necessary to measure arterial structure and function by MRI or CT. These methods also involve complicated techniques and take a long time, and CT results in exposure to radiation. Therefore, an easier screening tool is required at the clinical level.

NIRS can monitor the balance between oxygen consumption and oxygen supply by measuring optical absorption changes in oxygenated and deoxygenated Hb. A previous study [24] suggested that oxygenation as determined by NIRS has an influence on arterial function and severity of cardiovascular disease (CVD). It is therefore possible that such NIRS measurements could be used to evaluate an individual's risk of CVD.

4.3.1 Arterial Function

ASO is characterized by occlusive lesions consisting primarily of atheromas, which are often accompanied by fibrosis and calcification of the medial coat of the artery and that may be associated with thrombosis of varying severity. This disease is now considered to be an occlusive arterial disease

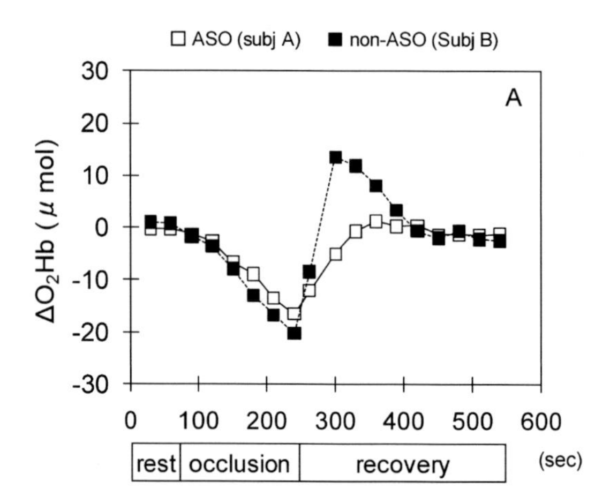

Fig. 4.3 Change in delta oxygenated Hb during the occlusion test: *Open squares* represents data for an arteriosclerosis obliterans (*ASO*) patient and *closed squares* that of a healthy subject (*non-ASO*). Note that compared with the healthy subject, a slower recovery after thigh occlusion was found in the gastrocnemius muscle of the arteriosclerosis obliterans patient [47]

predominantly affecting people with diabetes mellitus. ABI is the standard method for screening patients for ASO severity [43]. To evaluate basic physiological functions, however, rehabilitation practitioners require a simple method that is easy to follow and yields data critical to a diagnosis. Many researchers have attempted to evaluate or diagnose ASO severity using NIRS [44–47]. People with ASO have larger decreases in oxygenated Hb and blood volume during some tasks (walking/running, standing planter flexion, or occlusion), and they have slower recovery compared to healthy subjects. Figure 4.3 shows the typical changes in oxygenated Hb (ΔO_2Hb) in the gastrocnemius muscle during the thigh occlusion test. Compared to healthy subjects (non-ASO), the ASO patients have a slower recovery and a shallower recovery rate gradient after occlusion [47]. ABI correlated with both recovery time ($r = -0.61, p < 0.01$) and recovery rate ($r = 0.80, p < 0.001$). In the clinical setting, ASO patients have a faster recovery of their oxygenated Hb after taking medication to improve their circulation [46]. These patients reported reduced symptoms of ASO, including a greater ability to walk without pain, as determined by the subjective judgment of a medical doctor.

To evaluate endothelial function, the arterial structure must be measured, which can be carried out via the cold pressor test (CPT), which results in production of NO via sympathetic activation. This is a novel method for evaluating endothelial function [48]; however, it is also complicated and time consuming. The responses of oxygenation around the calf muscle and carotid artery diameter to a CPT have been measured simultaneously [49]. A greater decrease in oxygenated Hb and vasoconstriction during the test was found in elderly hypertensive patients compared with healthy elderly subjects. The response of oxygenated Hb was significantly correlated to vasodilation/vasoconstriction ($r = 0.658$, $p < 0.001$). These findings suggest that evaluation of oxygenation by NIRS might be able to provide valuable information about vascular endothelial function.

4.3.2 Microvascular Function

Assessment of microvascular compliance is a difficult task. Strain-gauge plethysmography, the standard method for evaluating vascular compliance, does not allow a distinction to be made between the venous and arterial blood volume shifts that occur for a given pressure change. Measurement of vascular compliance reflects the contribution of both compartments, even though in classic strain-gauge plethysmography the data obtained are attributable to only the dilation of the venous tree. Binzoni et al. [50] developed a novel method for measuring the compliance of the superficial veins of the lower limb by NIRS. This method is complementary to strain-gauge plethysmography. The NIRS data during head-up tilt allowed for assessment of superficial venous compliance of the lower limb at less than 24 mmHg of hydrostatic pressure (P), and also of changes in oxygenated and deoxygenated Hb levels in the P range -16 to 100 mmHg, which made it possible to assess the characteristics of the vasomotor response of the arteriolar system.

4.4 Brain Function

Similar to positron emission tomography (PET) and functional MRI (fMRI), NIRS can measure changes in cerebral blood flow and Hb saturation. A fast response to neuronal activation can be obtained with other related techniques, such as event-associated optical signals, which are derived from the near-infrared light scattering properties of active neurons. The NIRS method is used to measure brain oxygenation in adults [51], newborn infants [52], and the fetus during labor [53].

4.4.1 Vascular Dementia

The typical symptoms of subcortical vascular dementia (SVD) are slowness in motor performance and in cognitive processing, which results from early impairment of attention and executive functions [54]. Previous studies of SVD patients have observed a reduction in cerebral blood flow or the metabolic rate of oxygen during a resting or active state [55, 56]. In a recent study using simultaneous NIRS and fMRI [56], SVD patients showed decreased oxygenated Hb, total Hb, level-dependent blood concentration, and cerebral blood flow, and the metabolic rate of oxygen decreased during a simple motor task, which might explain the pathological small-vessel compromise, impaired vascular reactivity, and impaired neurovascular coupling. This NIRS method can therefore reveal various hemodynamic and metabolic changes, and may be useful in early detection or monitoring of subcortical vascular dementia.

4.4.2 Brain Oxygenation during Surgery

Many clinical studies and case reports have demonstrated that an ability to measure brain oxygenation aids in detecting clinically silent episodes of brain ischemia in a variety of clinical settings, thus helping to safeguard brain function. In patients with subarachnoid hemorrhage, episodes of angiographic cerebral vasospasm were associated with a reduction in ipsilateral NIRS signals [57]. Degree of spasm was associated with a greater reduction in same-side NIRS signal, demonstrating real-time detection of intracerebral ischemia.

The proper management of brain oxygenation is one of the principal endpoints of all anesthetic procedures, but the brain remains one of the last monitored organs during clinical anesthesiology. There are some medical procedures where iatrogenic brain ischemia is present, including carotid endarterectomy (CEA) in patients with high-grade carotid artery stenosis, temporary clipping in brain aneurysm surgery, hypothermic circulatory arrest for aortic arch procedures, and other cases in which the pathology itself generates brain ischemia, such as traumatic brain injury and stroke [58]. Proper monitoring of ischemia and response to treatment are essential for ensuring a good outcome.

There have been a number of case–control and retrospective studies of brain oximetry in cardiac surgeries that have shown improvements in outcome associated with brain oximetry monitoring and correlations between desaturation and adverse outcomes [59]. In a study utilizing brain oximetry during coronary artery bypass surgery where patients were randomized to active monitoring and a series of interventions designed to improve oxygen saturation, a significant association was found between brain desaturation and early cognitive decline, and also an increased risk of a prolonged hospital stay [60].

Various studies have shown that cerebral oximetry monitoring during CEA can be a valuable tool for detecting cerebral ischemia [61, 62]. A large cohort of NIRS findings from CEA performed under general anesthesia was studied to determine the sensitivity, specificity, and predictive values of regional cerebral tissue oxygen saturation cut-off points in predicting the need for shunting or for levels resulting in neurological complications [63]. In comparison with other devices, NIRS can provide continuous measurement of frontal cortex oxygenation in a simple and noninvasive way. Most studies utilizing NIRS during CEA have defined the sensitivity and specificity of changes in brain desaturation as correlated with either clinical signs of brain ischemia or other neuromonitoring modalities. Kirkpatrick et al. [64] reported a positive correlation between changes in cerebral flow velocity as detected by transcranial Doppler and changes in Hb saturation.

4.4.3 Brain Oxygenation in Children

In such complex settings as pediatric cardiac surgery, pediatric neurosurgery, and pediatric/neonatal intensive care, NIRS has been used to monitor and detect episodes of cerebral ischemia intraoperatively when combined with bispectral index monitoring [65]. NIRS monitors have been continuously and safely employed on the many neonates who suffer from dysfunction in brain oxygenation. Optical absorption by Hb provides a natural contrast agent to study oxygenation and dynamic processes, including evoked responses. The fact that the neonatal head is relatively transparent to light has been known for a long time, and transillumination of the head has been widely used to observe abnormalities near the brain surface, including development of hydrocephalus and subdural hemorrhage [2].

Brain oximetry has been employed to evaluate variations in the cerebral circulation in preterm infants presenting with apneic episodes [66]. An increased total Hb and decreased oxygen saturation were found during apneic episodes. In neonatal birth asphyxia, mild brain cooling has been utilized in an attempt to minimize subsequent brain hyperanemia and intravenous hyperalimentation. Ancora et al. [67] reported changes in brain oxygenation and EEG in an asphyxiated newborn during and after cool cap treatment, and suggested that NIRS might be useful during hypothermia treatments in order to monitor changes in brain oxygenation as possible indicators of the efficacy of cool cap treatment.

These findings suggest that the use of NIRS in monitoring brain oxygenation can provide indications of compromised brain perfusion in a clinical setting. The simple operation and continuous nature of the NIRS method must be considered with regard to the method's relative sensitivity and specificity versus other monitoring modalities.

4.5 Cancer Detection

There are many medical devices available to clinicians who diagnose, stage, and treat human cancer: CT, ultrasound, single-photon emission computed tomography, PET, MRI, and NIRS [68]. The NIRS method has been used more often in recent times. Its use NIRS as a supplement to conventional techniques in clinical areas has generated considerable interest. NIRS has such advantages as being noninvasive, fast, and relatively inexpensive. In addition, it poses no risk of ionizing radiation, and NIR light can easily penetrate centimeter-thick tissues. Differences in the NIRS signals between tissues are manifestations of multiple physiological changes, which are in turn associated with such factors as vascularization, cellularity, oxygen consumption, or remodeling [69].

The NIRS method offers the promise of a dramatic improvement in screening, and will continue to improve over the next decade. Diagnostically significant results in terms of sensitivity and specificity have been reported for breast, skin, pancreatic, and colorectal cancers. An accuracy between 72% and 97% has been achieved for the different types of tumors studied [70].

4.5.1 Breast Tissue

X-ray mammography is widely used in screening patients for breast cancer, the most common form of carcinoma in females. Because of uncertainties associated with radiographic density, mammography has an up to 22% false-negative rate, as well as a high false-positive rate in patients under 50 years of age [71]. A recent study found that routine initial mammography was not clinically advantageous for subjects under 35 years of age [72]. Furthermore, the use of hormone replacement therapy in postmenopausal patients is known to increase mammographic density [73] and has recently been shown to impede the efficacy of mammographic screening [74]. Techniques such as MRI and ultrasound are used only as secondary procedures because of such factors as high cost and poor specificity or low sensitivity.

NIRS methods are advantageous for screening because they are noninvasive, quantitative, and relatively inexpensive. They do not require compression and pose no risk of ionizing radiation. Chance's group [75] introduced the endogenous contrast method, tomographic near-infrared imaging of the human breast, in 1994 [75]. With this technique the inherent properties of normal and malignant tissue can be examined noninvasively using invisible near-infrared light. Recent improvements include the development of a handheld device for breast cancer detection [76] and the use of MRI to improve 3D reconstruction of optical images [77]. Diffuse optical spectroscopy by increasing the wavelength range and resolution has also been employed to characterize malignant tumors (Fig. 4.4) [78].

The use of invasive procedures such as fine-needle aspiration or surgical biopsy have also been implemented to provide a definitive diagnosis for breast cancer. Given the suboptimal performance of x-ray mammography in premenopausal and perimenopausal patients, the majority of invasive follow-up procedures are performed on normal or benign tissue that present no malignant disease [79]. As a result, the use of noninvasive NIR methods as a supplement to present techniques for diagnosing and detecting breast cancer has generated considerable interest.

4.5.2 Skin Lesion

Skin cancer is one the most common human carcinomas. The clinical diagnosis is often difficult, as many benign skin lesions resemble malignancies upon visual examination. Therefore, a histopathologic analysis of skin biopsies remains the standard for confirmation. McIntosh et al. [80] collected visible and near-infrared spectra of skin neoplasms in vivo and found significant differences between normal skin and skin lesions in several areas. NIRS is therefore a promising noninvasive technique for screening of skin lesions.

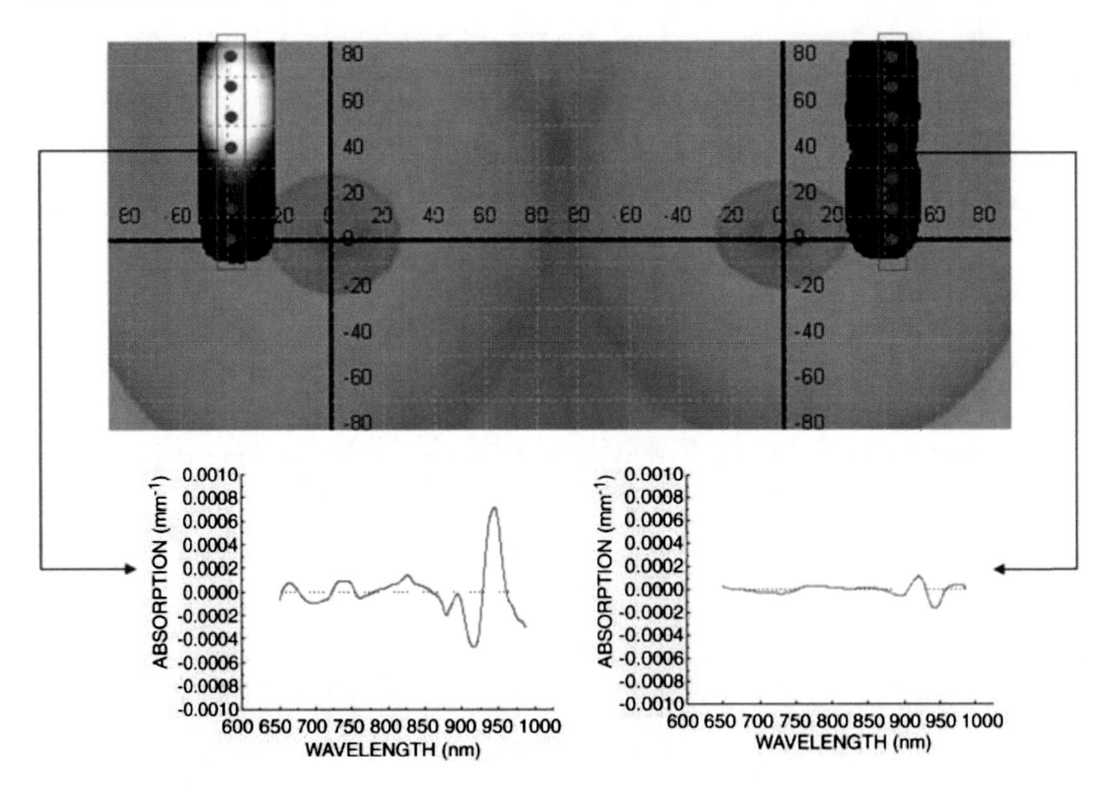

Fig. 4.4 Specific tumor component index-based image of a malignant tumor: Diffuse optical spectroscopic imaging-mapped regions of the breast are superimposed onto a breast picture. The *vertical boxes* indicate a subset of the region of interest corresponding to the spectra below. *Dots* on the image indicate locations at which imaging measurements were obtained. Spectra are shown in two regions: malignant (*left*) and normal tissue (*right*). Note that the specific tumor component absorption spectrum is found only in the tumor-containing region and not in surrounding normal tissue [78]

4.5.3 Therapeutic Application

Various treatments are used to cure cancer, control its growth, shrink tumors, or to destroy microscopic cancer cells. Chemotherapy, and radiation/photon dynamic therapy are two of the most often-used procedures in cancer treatment. NIRS can offer options for optimizing the biological effect, accurate dosimetry, and for monitoring treatment progression and efficacy. In a previous study [69], the tissue optical index, developed as a contrast function by combining diffuse optical spectroscopy (DOS) measurements, was monitored following a single dose of adriamycin and cyclophosphamide neoadjuvant chemotherapy. DOS can be used as a quantitative diffuse optical method, which is conceptually similar to the relationship between MRI and MRS. In general, DOS employs a limited number of source-detector positions; however, it employs broadband content in the temporal and spectral domains in order to recover complete absorption and scattering spectra measuring approximately 650–1,000 nm [69]. Figure 4.5 shows the tissue oxygen index (TOI) prior to and on days 1, 2, 3, 6, and 8 following therapy. Note the dramatic drop in TOI within just 1 day: 2.5 prior to therapy to 1.7 on day 1. By day 8, peak TOI levels were approximately equal to normal baseline levels, thus representing a 60% reduction in 1 week. NIRS can therefore be used to quantitatively assess the biochemical composition of the tumor and its response to treatment.

From these findings it can be concluded that NIRS may play an important role in detecting cancer, predicting responders early in the course of therapy, and developing an optimal strategy for care.

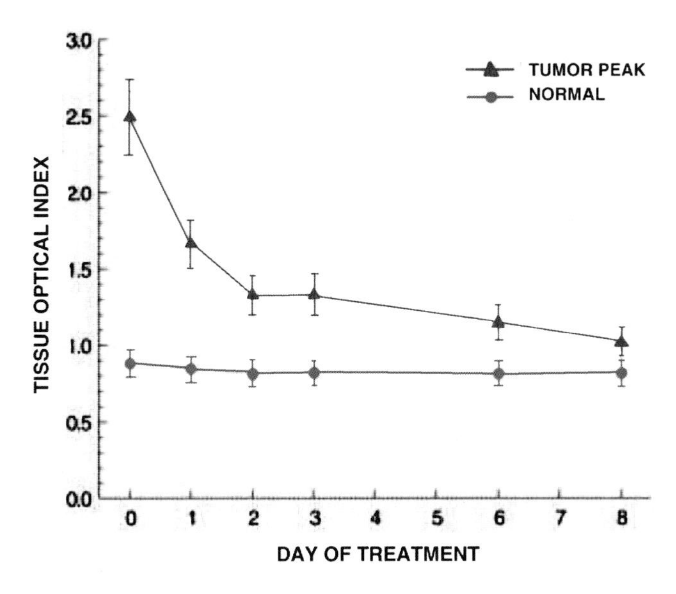

Fig. 4.5 Changes in tissue oxygen index observed post-therapy: Time point 0 was prior to treatment. The error bars represent the standard deviation of measurements. Note that changes were observed in the tissue oxygen index of the tumor (*triangles*) as soon as 1 day post-therapy [69]

4.6 Others

Although it has already been applied to muscle/brain metabolism, vessel function, and cancer detection, NIRS is also quite useful in evaluating and diagnosing various diseases in the clinical setting.

Preterm labor is responsible for 70% of perinatal mortality and nearly half of long-term neurological morbidity [81], and is one of the most challenging problems in modern obstetrics and gynecology. One of the keys to treating preterm labor and preventing severe neonatal complications is early identification of at-risk patients. Hornung et al. [82] used frequency-domain NIRS to monitor changes in the optical properties of the uterine cervix that occur during regular pregnancies, and thus showed that oxygenated Hb and total Hb are significantly correlated with gestational age. Therefore, this kind of NIRS device can be applied to assess the cervix for prediction of risk of preterm labor or for distinguishing between true and false labor.

Surgical site infections (SSIs) comprise the most common serious complication of anesthesia and surgery. The transition from contamination to an established infection occurs during a decisive period, even though infections are typically detected a week or more following surgery [83]. Antibiotics administered during this decisive period are more effective in reducing infection risk. Previous studies have suggested that tissue oxygen tension is one of the best factors for predicting SSIs [84, 85]. Govinda et al. [86] measured changes in tissue oxygen saturation (StO_2) as determined by NIRS after colorectal surgery in order to establish the parameters for predicting development of an SSI. StO_2 measured at the upper arm was found to be lower in patients with an SSI than in those without, and these measurements had a high sensitivity and specificity for predicting development of an SSI. NIRS may thus be able to predict surgical site infections and so allow earlier preventive or treatment measures to be implemented.

Severe sepsis is an inflammatory response to infection that results in acute organ dysfunction [87]. Despite a hyperdynamic systemic blood flow and elevated mixed venous oxygen saturations, elevated blood lactate concentrations are common in sepsis and suggest that a maldistribution of oxygen within the tissues may contribute to organ failure and death. In human sepsis, impaired microvascular reactivity is related to tissue dysoxia as well as organ dysfunction. Doerschug et al. [88] measured microvascular Hb and StO_2 using NIRS both before and after forearm ischemia in patients with sepsis and found StO_2 to decrease and the rate of tissue reoxygenation to be impaired with sepsis. Additionally, the rate of tissue reoxygenation was shown to be related to degree of organ dysfunction. NIRS measurements of tissue microvascular perfusion and reactivity may provide important information about sepsis, and such information may serve as endpoints in future therapeutic interventions aimed at improving microvascularization [88].

4.7 Summary

Improvements in NIRS during the past eight decades have matured this technique to the point where it is now effective in a wide range of applications.

However, most of the NIRS investigations reported thus far are feasibility studies. In order to become a major part of a monitoring system, more technological improvements will be needed. Further studies with a larger number of patients and/or normal subjects are also required to further understand the problems associated with this technique, and to confirm its efficacy. Such clinical trials are expected to be carried out over the next decade and will likely have an impact in both research and clinical settings. In sum, NIRS is a prospective technology, an d this kind of tool might play a crucial role in diagnosing and treating human disease, or for defining response to therapy.

Problem

4.1 How does NIRS help us better understand the relationship between vasculature and tissue function?

Further Reading

Boushel R, Langberg H, Olesen J, Gonzales-Alonzo J, Bülow J, Kjaer M (2001) Monitoring tissue oxygen availability with near infrared spectroscopy (NIRS) in health and disease. Scand J Med Sci Sports 11(4):213–222

Hebden JC, Delpy DT (1997) Diagnostic imaging with light. Br J Radiol 70:S206–S214

Kondepati VR, Heise HM, Backhaus J (2008) Recent applications of near-infrared spectroscopy in cancer diagnosis and therapy. Anal Bioanal Chem 390(1):125–139

Murkin JM, Arango M (2009) Near-infrared spectroscopy as an index of brain and tissue oxygenation. Br J Anaesth 103(Suppl 1):i3–i13

References

1. Jöbsis FF (1977) Non-invasive infrared monitoring of cerebral and myocardial oxygen sufficiency and circulatory parameters. Science 198:1264–1267
2. Hebden JC, Delpy DT (1997) Diagnostic imaging with light. Br J Radiol 70:S206–S214

3. Watkin SL, Spencer SA, Dimmock PW, Wickramasinghe YA, Rolfe P (1999) A comparison of pulse oximetry and near infrared spectroscopy (NIRS) in the detection of hypoxaemia occurring with pauses in nasal airflow in neonates. J Clin Monit Comput 15(7–8):441–447
4. Chance B, Nioka S, Kent J, McCully K, Fountain M, Greenfeld R, Holtom G (1988) Time-resolved spectroscopy of hemoglobin and myoglobin in resting and ischemic muscle. Anal Biochem 174(2):698–707
5. Cope M, Delpy DT (1988) System for long-term measurement of cerebral blood and tissue oxygenation on newborn infants by near infrared transillumination. Med Biol Eng Comput 26(3):289–294
6. Suzuki S, Takasaki S, Ozaki T, Kobayashi Y (1999) Tissue oxygenation monitor using NIR spatially resolved spectroscopy. Proc SPIE 3597:582–592
7. Miwa M, Ueda Y, Chance B (1995) Development of time-resolved spectroscopy system for quantitative noninvasive tissue measurement. Proc SPIE 2389:142–149
8. Duncan A, Whitlock TL, Cope M, Delpy DT (1993) Multiwavelength, wideband, intensity-modulated optical spectrometer for near-infrared spectroscopy and imaging. Proc SPIE 1888:248–257
9. McCully KK, Landsberg L, Suarez M, Hofmann M, Posner JD (1997) Identification of peripheral vascular disease in elderly subjects using optical spectroscopy. J Gerontol A Biol Sci Med Sci 52(3):B159–B165
10. Boushel R, Langberg H, Olesen J, Gonzales-Alonzo J, Bülow J, Kjaer M, Scand J (2001) Monitoring tissue oxygen availability with near infrared spectroscopy (NIRS) in health and disease. Med Sci Sports 11(4):213–222
11. Hamaoka T, McCully K, Chance B, Iwane H (1994) Noninvasive measures of muscle metabolism. In: Sen C, Packer L, Hanninen O (eds) Exercise and oxygen toxicity. Elsevier Science, Amsterdam, pp 481–510
12. Cerretelli P, Binzoni T (1997) The contribution of NMR, NIRS and their combination to the functional assessment of human muscle. Int J Sports Med 18(Suppl 4):S270–S279
13. Tran TK, Sailasuta N, Kreutzer U, Hurd R, Chung Y, Mole P, Kuno S, Jue T (1999) Comparative analysis of NMR and NIRS measurements of intracellular PO_2 in human skeletal muscle. Am J Physiol 276(6 Pt 2):R1682–R1690
14. Miura H, Araki H, Matoba H, Kitagawa K (2000) Relationship among oxygenation, myoelectric activity, and lactic acid accumulation in vastus lateralis muscle during exercise with constant work rate. Int J Sports Med 21(3):180–184
15. Chance B, Dait MT, Zhang C, Hamaoka T, Hagerman F (1992) Recovery from exercise-induced desaturation in the quadriceps muscles of elite competitive rowers. Am J Physiol 262(3 Pt 1):C766–C775
16. Rundell KW, Nioka S, Chance B (1997) Hemoglobin/myoglobin desaturation during speed skating. Med Sci Sports Exerc 29(2):248–258
17. Shinohara M, Kouzaki M, Yoshihisa T, Fukunaga T (1998) Mechanomyogram from the different heads of the quadriceps muscle during incremental knee extension. Eur J Appl Physiol Occup Physiol 78(4):289–295
18. Colier WN, Meeuwsen IB, Degens H, Oeseburg B (1995) Determination of oxygen consumption in muscle during exercise using near-infrared spectroscopy. Acta Anaesthesiol Scand 107(Suppl):151–155
19. Bhambhani Y, Maikala R, Buckley S (1998) Muscle oxygenation during incremental arm and leg exercise in men and women. Eur J Appl Physiol Occup Physiol 78(5):422–431
20. Kahn JF, Jouanin JC, Bussière JL, Tinet E, Avrillier S, Ollivier JP, Monod H (1998) The isometric force that induces maximal surface muscle deoxygenation. Eur J Appl Physiol Occup Physiol 78(2):183–187
21. Boushel R, Pott F, Madsen P, Rådegran G, Nowak M, Quistorff B, Secher N (1998) Muscle metabolism from near-infrared spectroscopy during rhythmic handgrip in humans. Eur J Appl Physiol Occup Physiol 79(1):41–48
22. McCully KK, Iotti S, Kendrick K, Wang Z, Posner JD, Leigh J Jr, Chance B (1994) Simultaneous in vivo measurements of HbO_2 saturation and PCr kinetics after exercise in normal humans. J Appl Physiol 77(1):5–10
23. Miura H, McCully K, Hong L, Nioka S, Chance B (2001) Regional difference of muscle oxygen saturation and blood volume during exercise determined by near infrared imaging device. Jpn J Physiol 51(5):599–606
24. Miura H, McCully K, Chance B (2003) Application of multiple NIRS imaging device to the exercising muscle metabolism. Spectroscopy 17:549–558
25. Miura H, McCully K, Nioka S, Chance B (2004) Relationship between muscle architectural features and oxygenation status determined by near infrared device. Eur J Appl Physiol 91(2–3):273–278
26. Moritani T, Nagata A, Muro M (1982) Electromyographic manifestations of muscular fatigue. Med Sci Sports Exerc 14(3):198–202
27. Puente-Maestu L, Tena T, Trascasa C, Pérez-Parra J, Godoy R, García MJ, Stringer WW (2003) Training improves muscle oxidative capacity and oxygenation recovery kinetics in patients with chronic obstructive pulmonary disease. Eur J Appl Physiol 88(6):580–587
28. Ichimura S, Murase N, Osada T, Kime R, Homma T, Ueda C, Nagasawa T, Motobe M, Hamaoka T, Katsumura T (2006) Age and activity status affect muscle reoxygenation time after maximal cycling exercise. Med Sci Sports Exerc 38(7):1277–1281
29. Kime R, Karlsen T, Nioka S, Lech G, Madsen O, Sæterdal R, Im J, Chance B, Stray-Gundersen J (2003) Discrepancy between cardiorespiratory system and skeletal muscle in elite cyclists after hypoxic training. Dyn Med 2(1):4

30. Bhambhani Y, Tuchak C, Burnham R, Jeon J, Maikala R (2000) Quadriceps muscle deoxygenation during functional electrical stimulation in adults with spinal cord injury. Spinal Cord 38(10):630–638
31. Aldayel A, Muthalib M, Jubeau M, McGuigan M, Nosaka K (2010) Muscle oxygenation of vastus lateralis and medialis muscles during alternating and pulsed current electrical stimulation. Eur J Appl Physiol 111(5):779–787
32. Boushel R, Langberg H, Olesen J, Gonzales-Alonzo J, Bülow J, Kjaer M (2001) Monitoring tissue oxygen availability with near infrared spectroscopy (NIRS) in health and disease. Scand J Med Sci Sports 11(4):213–222
33. Wilson JR, Mancini DM, McCully K, Ferraro N, Lanoce V, Chance B (1989) Noninvasive detection of skeletal muscle underperfusion with near-infrared spectroscopy in patients with heart failure. Circulation 80(6):1668–1674
34. Mancini DM, Wilson JR, Bolinger L, Li H, Kendrick K, Chance B, Leigh JS (1994) In vivo magnetic resonance spectroscopy measurement of deoxymyoglobin during exercise in patients with heart failure: demonstration of abnormal muscle metabolism despite adequate oxygenation. Circulation 90(1):500–508
35. Wariar R, Gaffke JN, Haller RG, Bertocci LA (2000) A modular NIRS system for clinical measurement of impaired skeletal muscle oxygenation. J Appl Physiol 88(1):315–325
36. Mancini DM, Ferraro N, Nazzaro D, Chance B, Wilson JR (1991) Respiratory muscle deoxygenation during exercise in patients with heart failure demonstrated with near-infrared spectroscopy. J Am Coll Cardiol 8 (2):492–498
37. Mancini DM, La Manca J, Donchez L, Henson D, Levine S (1996) The sensation of dyspnea during exercise is not determined by the work of breathing in patients with heart failure. J Am Coll Cardiol 28(2):391–395
38. Matsen FA 3rd, Winquist RA, Krugmire RB Jr (1980) Diagnosis and management of compartmental syndromes. J Bone Joint Surg Am 62(2):286–291
39. Arató E, Kürthy M, Sínay L, Kasza G, Menyhei G, Masoud S, Bertalan A, Verzár Z, Kollár L, Roth E, Jancsó G (2009) Pathology and diagnostic options of lower limb compartment syndrome. Clin Hemorheol Microcirc 41(1):1–8
40. Corretti MC, Plotnick GD, Vogel RA (1995) The effects of age and gender on brachial artery endothelium-dependent vasoactivity are stimulus-dependent. Clin Cardiol 18(8):471–476
41. Najjar SS, Scuteri A, Lakatta EG (2005) Arterial aging: is it an immutable cardiovascular risk factor? Hypertension 46(3):454–462
42. Cameron JD, Dart AM (1994) Exercise training increases total systemic arterial compliance in humans. Am J Physiol 266(2 Pt 2):H693–H701
43. Weitz JI, Byrne J, Clagett GP, Farkouh ME, Porter JM, Sackett DL, Strandness DE Jr, Taylor LM (1996) Diagnosis and treatment of chronic arterial insufficiency of the lower extremities: a critical review. Circulation 94(11):3026–3049
44. Komiyama T, Shigematsu H, Yasuhara H, Muto T (1994) An objective assessment of intermittent claudication by near-infrared spectroscopy. Eur J Vasc Surg 8(3):294–296
45. Kooijman HM, Hopman MT, Colier WN, van der Vliet JA, Oeseburg B (1997) Near infrared spectroscopy for noninvasive assessment of claudication. J Surg Res 72(1):1–7
46. Miura H, McCully K, Hong L, Nioka S, Chance B (2000) Exercise-induced changes in oxygen status in calf muscle of elderly subjects with peripheral vascular disease using functional near infrared imaging machine. Ther Res 21(6):79–84
47. Miura H, Okumura N (2010) A novel approach to evaluate the vessel function determined by near infrared spectroscopy. Adv Exp Med Biol 662:467–471
48. Tousoulis D, Davies G, Tentolouris C, Crake T, Toutouzas P (1997) Inhibition of nitric oxide synthesis during the cold pressor test in patients with coronary artery disease. Am J Cardiol 79(12):1676–1679
49. Miura H, Takahashi Y, Okumura N (2011) Response of peripheral vascular system to cold pressor test measured by near infrared spectroscopy. J Jpn Coll Angiol 51:255–257
50. Binzoni T, Quaresima V, Ferrari M, Hiltbrand E, Cerretelli P (2000) Human calf microvascular compliance measured by near-infrared spectroscopy. J Appl Physiol 88(2):369–372
51. Wyatt JS (1993) Near-infrared spectroscopy in asphyxial brain injury. Clin Perinatol 20(2):369–378
52. Elwell CE, Owen-Reece H, Cope M, Wyatt JS, Edwards AD, Delpy DT, Reynolds EO (1993) Measurement of adult cerebral haemodynamics using near infrared spectroscopy. Acta Neurochir Suppl 59:74–80
53. Wyatt JS, Peebles DM (1993) Near infrared spectroscopy and intrapartum fetal surveillance. In: Spencer JAD (ed) Intrapartum fetal surveillance. RCOG Press, London, pp 329–345
54. Román GC (1987) Senile dementia of the Binswanger type: a vascular form of dementia in the elderly. JAMA 258(13):1782–1788
55. Yoshikawa T, Murase K, Oku N, Kitagawa K, Imaizumi M, Takasawa M, Nishikawa T, Matsumoto M, Hatazawa J, Hori M (2003) Statistical image analysis of cerebral blood flow in vascular dementia with small-vessel disease. J Nucl Med 44(4):505–511
56. Tak S, Yoon SJ, Jang J, Yoo K, Jeong Y, Ye JC (2011) Quantitative analysis of hemodynamic and metabolic changes in subcortical vascular dementia using simultaneous near-infrared spectroscopy and fMRI measurements. Neuroimage 55(1):176–184

57. Bhatia R, Hampton T, Malde S, Kandala NB, Muammar M, Deasy N, Strong A (2007) The application of near-infrared oximetry to cerebral monitoring during aneurysm embolization: a comparison with intraprocedural angiography. J Neurosurg Anesthesiol 19(2):97–104
58. Murkin JM, Arango M (2009) Near-infrared spectroscopy as an index of brain and tissue oxygenation. Br J Anaesth 103(Suppl 1):i3–i13
59. Edmonds HL Jr, Ganzel BL, Austin EH III (2004) Cerebral oximetry for cardiac and vascular surgery. Semin Cardiothorac Vasc Anesth 8(2):147–166
60. Slater JP, Guarino T, Stack J, Vinod K, Bustami RT, Brown JM 3rd, Rodriguez AL, Magovern CJ, Zaubler T, Freundlich K, Parr GV (2009) Cerebral oxygen desaturation predicts cognitive decline and longer hospital stay after cardiac surgery. Ann Thorac Surg 87(1):36–45
61. Moritz S, Kasprzak P, Arlt M, Taeger K, Metz C (2007) Accuracy of cerebral monitoring in detecting cerebral ischemia during carotid endarterectomy: a comparison of transcranial Doppler sonography, near-infrared spectroscopy, stump pressure, and somatosensory evoked potentials. Anesthesiology 107(4):563–569
62. Yamamoto K, Miyata T, Nagawa H (2007) Good correlation between cerebral oxygenation measured using near infrared spectroscopy and stump pressure during carotid clamping. Int Angiol 26(3):262–265
63. Mille T, Tachimiri ME, Klersy C, Ticozzelli G, Bellinzona G, Blangetti I, Pirrelli S, Lovotti M, Odero A (2004) Near infrared spectroscopy monitoring during carotid endarterectomy: which threshold value is critical? Eur J Vasc Endovasc Surg 27(6):646–650
64. Kirkpatrick PJ, Lam J, Al-Rawi P, Smielewski P, Czosnyka M (1998) Defining thresholds for critical ischemia by using near-infrared spectroscopy in the adult brain. J Neurosurg 89(3):389–394
65. Hayashida M, Kin N, Tomioka T, Orii R, Sekiyama H, Usui H, Chinzei M, Hanaoka K (2004) Cerebral ischaemia during cardiac surgery in children detected by combined monitoring of BIS and near-infrared spectroscopy. Br J Anaesth 92(5):662–669
66. Yamamoto A, Yokoyama N, Yonetani M, Uetani Y, Nakamura H, Nakao H (2003) Evaluation of change of cerebral circulation by SpO_2 in preterm infants with apneic episodes using near infrared spectroscopy. Pediatr Int 45(6):661–664
67. Ancora G, Maranella E, Locatelli C, Pierantoni L, Faldella G (2009) Changes in cerebral hemodynamics and amplitude integrated EEG in an asphyxiated newborn during and after cool cap treatment. Brain Dev 31(6):442–444
68. Frangioni JV (2008) New technologies for human cancer imaging. J Clin Oncol 26(24):4012–4021
69. Tromberg BJ, Cerussi A, Shah N, Compton M, Durkin A, Hsiang D, Butler J, Mehta R (2005) Imaging in breast cancer: diffuse optics in breast cancer: detecting tumors in pre-menopausal women and monitoring neoadjuvant chemotherapy. Breast Cancer Res 7(6):279–285
70. Kondepati VR, Heise HM, Backhaus J (2008) Recent applications of near-infrared spectroscopy in cancer diagnosis and therapy. Anal Bioanal Chem 390(1):125–139
71. Kerlikowske K, Barclay J (1997) Outcomes of modern screening mammography. J Natl Cancer Inst Monogr 22:105–111
72. Hindle WH, Davis L, Wright D (1999) Clinical value of mammography for symptomatic women 35 years of age and younger. Am J Obstet Gynecol 180(6 Pt 1):1484–1490
73. Baines CJ, Dayan R (1999) A tangled web: factors likely to affect the efficacy of screening mammography. J Natl Cancer Inst 91(10):833–838
74. Laya MB, Larson EB, Taplin SH, White E (1996) Effect of estrogen replacement therapy on the specificity and sensitivity of screening mammography. J Natl Cancer Inst 88(10):643–649
75. Nioka S, Miwa M, Orel S, Shnall M, Haida M, Zhao S, Chance B (1994) Optical imaging of human breast cancer. Adv Exp Med Biol 361:171–179
76. Hsiang D, Shah N, Yu H, Su MY, Cerussi A, Butler J, Baick C, Mehta R, Nalcioglu O, Tromberg B (2005) Coregistration of dynamic contrast enhanced MRI and broadband diffuse optical spectroscopy for characterizing breast cancer. Technol Cancer Res Treat 4(5):549–558
77. Carpenter CM, Pogue BW, Jiang S, Dehghani H, Wang X, Paulsen KD, Wells WA, Forero J, Kogel C, Weaver JB, Poplack SP, Kaufman PA (2007) Image-guided optical spectroscopy provides molecular-specific information in vivo: MRI-guided spectroscopy of breast cancer hemoglobin, water, and scatterer size. Opt Lett 32(8):933–935
78. Kukreti S, Cerussi AE, Tanamai W, Hsiang D, Tromberg BJ, Gratton E (2010) Characterization of metabolic differences between benign and malignant tumors: high-spectral-resolution diffuse optical spectroscopy. Radiology 254(1):277–284
79. Fletcher SW (1997) Breast cancer screening among women in their forties: an overview of the issues. J Natl Cancer Inst Monogr 22:5–9
80. McIntosh LM, Summers R, Jackson M, Mantsch HH, Mansfield JR, Howlett M, Crowson AN, Toole JW (2001) Towards non-invasive screening of skin lesions by near-infrared spectroscopy. J Invest Dermatol 116(1):175–181
81. Goldenberg RL, Hauth JC, Andrews WW (2000) Intrauterine infection and preterm delivery. N Engl J Med 342(20):1500–1507

82. Hornung R, Spichtig S, Baños A, Stahel M, Zimmermann R, Wolf M (2011) Frequency-domain near-infrared spectroscopy of the uterine cervix during regular pregnancies. Lasers Med Sci 26(2):205–212
83. Polk HC Jr (1974) The prophylaxis of infection following operative procedures. J Ky Med Assoc 72(3):139–143
84. Hopf HW, Hunt TK, West JM, Blomquist P, Goodson WH 3rd, Jensen JA, Jonsson K, Paty PB, Rabkin JM, Upton RA, von Smitten K, Whitney JD (1997) Wound tissue oxygen tension predicts the risk of wound infection in surgical patients. Arch Surg 132(9):997–1004
85. Knighton DR, Halliday B, Hunt TK (1984) Oxygen as an antibiotic: the effect of inspired oxygen on infection. Arch Surg 119(2):199–204
86. Govinda R, Kasuya Y, Bala E, Mahboobi R, Devarajan J, Sessler DI, Akça O (2010) Early postoperative subcutaneous tissue oxygen predicts surgical site infection. Anesth Analg 111(4):946–952
87. Bone RC, Balk RA, Cerra FB, Dellinger RP, Fein AM, Knaus WA, Schein RM, Sibbald WJ (1992) Definitions for sepsis and organ failure and guidelines for the use of innovative therapies in sepsis: the ACCP/SCCM Consensus Conference Committee. American College of Chest Physicians/Society of Critical Care Medicine. Chest 101(6):1644–1655
88. Doerschug KC, Delsing AS, Schmidt GA, Haynes WG (2007) Impairments in microvascular reactivity are related to organ failure in human sepsis. Am J Physiol Heart Circ Physiol 293(2):H1065–H1071

In-Vivo NIRS and Muscle Oxidative Metabolism

5

Takafumi Hamaoka

5.1 Introduction

Muscle oxidative metabolism has often in the past been examined by traditional analytical biochemistry based on obtaining biopsy samples. Invasive techniques such as intravascular catheterization have provided information on muscle oxygen status at rest and during exercise [1]. Myoglobin O_2 saturation and reduced nicotinamide adenine dinucleotide (NADH) redox state can be detected using freeze-clamped tissue [2]. The disadvantages of the invasive approach include poor time resolution due to repeated measurements and, of course, the invasive nature of the sampling method, so there is a need for noninvasive approaches to measure muscle oxygen status during exercise. NIRS has been utilized to evaluate skeletal muscle O_2 dynamics and energetics during exercise. Application of this technology has focused on the validity and calibration of measurements, and biochemical, physiological, and pathological research of muscle oxidative metabolism [3–6]. The primary reason why NIRS methodology is useful for evaluating muscle is the visibility of heme molecules and their oxygen-dependent characteristics.

Visible light has been used to monitor changes in tissue oxygenation since the 1930s [7]. Mitochondrial NADH signal measured using optical methods has shown a rapid change caused by electrical muscle stimulation (within a fraction of a second), indicating coupling of muscle contraction and mitochondrial activity [8, 9]. Jöbsis [10] then discovered that NIR light easily penetrates the skull, which set the stage for recent application of NIRS to scientific research in brain and muscle as well as in various clinical settings. A muscle NIRS system was then developed [11–13] that served as one of the first models to provide an opportunity for noninvasive and portable clinical muscle research.

5.2 Biochemistry and Physiology of Muscle Oxidative Metabolism

Skeletal muscle employs three major biochemical processes to synthesize ATP: oxidative phosphorylation (32–36 ATPs), anaerobic glycolysis (2 ATPs), and the PCr pathway (1 ATP). During low- to moderate-intensity exercise, like that involved in normal daily activities, skeletal muscle intensively

T. Hamaoka, M.D., Ph.D. (✉)
Department of Sport and Health Science, Ritsumeikan University, 1 Nojihigashi, Kusatsu, Shiga, Japan
e-mail: kyp02504@nifty.com

T. Jue and K. Masuda (eds.), *Application of Near Infrared Spectroscopy in Biomedicine*,
Handbook of Modern Biophysics 4, DOI 10.1007/978-1-4614-6252-1_5,

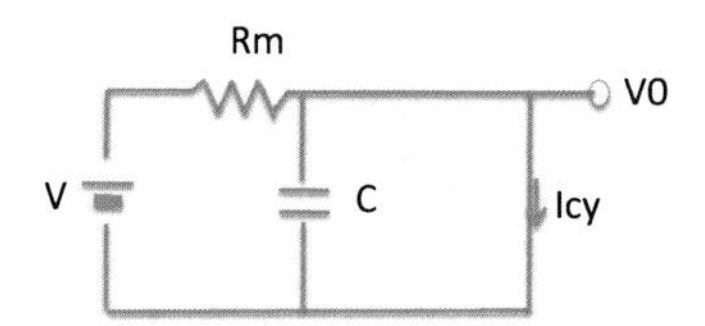

Fig. 5.1 Electrical analog model with a first-order linear system for respiratory control in muscle at submaximal oxidative rates. The current Icy represents the cytosolic adenosinetriphosphatase (ATPase) rate, the voltage V0 represents the free energy potential available in the mitochondria. Capacitance C is due to the creatine kinase reaction, and PCr is analogous to stored charge on a capacitor. Resistor Rm is a function of the number and properties of the mitochondria in the cell, and current Irm is the rate of oxidative phosphorylation [18]

relies on oxidative metabolism. Skeletal muscle O_2 consumption (VO_2) can be elevated 50-fold with up to tenfold abrupt increases in O_2 delivery (DO_2). The net oxidative energy pathway in muscles can be described by the following equation [14]:

$$3\mathrm{ADP} + 3\mathrm{Pi} + \mathrm{NADH} + \mathrm{H}^+ + {}^1\!/_2\mathrm{O}_2 = 3\mathrm{ATP} + \mathrm{NAD}^+ + \mathrm{H_2O}, \tag{5.1}$$

where ADP is adenosine diphosphate, Pi is inorganic phosphate, ATP is adenosine triphosphate, and NAD^+ is the nicotinamide adenine dinucleotide.

The kinetic control model of respiration that describes metabolic rate as a function of regulatory substrate concentrations using the Michaelis-Menten equation is written as [15]

$$V/V_\mathrm{m} = 1/(1 + k_1/\mathrm{ADP} + k_2/\mathrm{Pi} + k_3/\mathrm{O}_2 + k_4/\mathrm{NADH}), \tag{5.2}$$

where V is observed velocity, V_m is maximum velocity, k_{1-4} represent affinity constants for various substrates, Pi is inorganic phosphate, PCr is phosphocreatine, and NAD is the nicotinamide adenine dinucleotide. The in-vivo mitochondrial concentrations of ADP, Pi, O_2, and NADH are around 20, 1,000, 1, and 100 μM, respectively, and the in-vitro K_m (half-maximum velocity) values for ADP, Pi, O_2 (cytochrome aa3 for O_2), and NADH are 20, 300, 0.1, and ~10 μM, respectively. The in-vitro data indicate that the primary candidate for metabolic control is ADP [15]. However, it should be noted that no final agreement has been reached as to what is the metabolite controller of mitochondrial respiration.

For the phosphorylation cycle to function, ADP and Pi must be transported into the mitochondrial matrix, and then ATP is transported from there back to the cytoplasm. This reaction is catalyzed by the adenine nucleotide translocase of the inner mitochondrial membrane. It has thus been proposed that the rate of mitochondrial respiration can be determined by the rate of adenine nucleotide translocation and that, therefore, the [ATP]/[ADP] ratio regulates the respiratory rate under physiological conditions [16]. Holian et al. [17] generalized metabolic regulation to include ATP as a controller, as well as ADP and inorganic phosphate. The near-equilibrium hypothesis states that there is a correlation among the cellular [ATP]/[ADP][Pi] ratio, the mitochondrial [NAD]/[NADH] ratio, and respiration rate. Respiration rate would thus be dependent on [ATP]/[ADP][Pi] under conditions of Pi (~1 mM) and ATP (~6 mM). In general, ATP is maintained constant during submaximal exercise by creatine kinase equilibrium via the creatine shuttle in muscle, and ATP is not operative in respiration regulation.

The thermodynamic regulation model [18], an analogue of the electrical model for respiration control, in which PCr can be directly related to respiration rate, should thus be applicable to mitochondrial respiration (see Fig. 5.1). The current (I_{cy}) represents the cytosolic adenosine triphosphatase (ATPase) rate, and voltage V_0 represents the free energy potential available in the mitochondria. Capacitance C is due to the creatine kinase reaction, and PCr is analogous to the stored

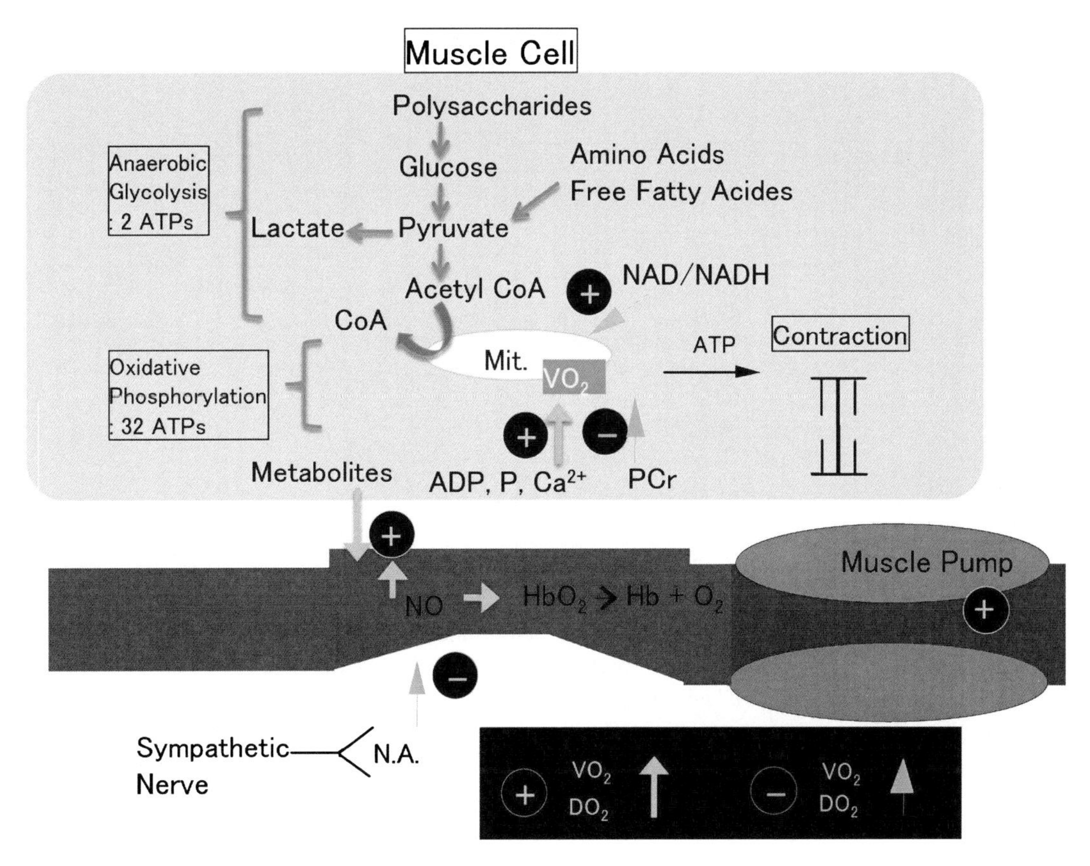

Fig. 5.2 Cellular mechanism for mitochondrial respiration and oxygen delivery to muscle. Mitochondrial respiration is controlled by the concentrations of ADP, Pi, and Ca^{2+}, the NAD/NADH ratio, and so on. Oxygen delivery (blood flow) is controlled by sympathetic vasoconstriction, varying metabolites, NO, muscle pump, etc. ATP is synthesized via the three main steps – namely, oxidative phosphorylation (32 ATPs), anaerobic glycolysis (2 ATPs), and PCr pathway (1 ATP). *Mit* mitochondrion, *NA* noradrenaline, *NO* nitric oxide: HbO_2 oxyhemoglobin, *ATP* adenosine triphosphate, *ADP* adenosine diphosphate, *Pi* inorganic phosphate, *PCr* Phosphocreatine, *NAD* nicotinamide adenine dinucleotide, *NADH* reduced nicotinamide adenine dinucleotide, VO_2 oxygen consumption: DO_2, oxygen delivery

charge on a capacitor. Resistor R_m is a function of the number and properties of the mitochondria in the cell, and current I_{rm} is the rate of oxidative phosphorylation [18], where $I_{rm} = I_c + I_{cy}$ and $I_c = C\mathrm{d}V_0/\mathrm{d}t$. The free energy of ATP hydrolysis (ΔG_{ATP}) becomes a function of ln{[PCr]/(Total-Cr – [PCr])^2} (see [18] for details). To a good approximation, ln{[PCr]/(Total-Cr – [PCr])^2} is linear for PCr between 20% and 70% of total-Cr. The cellular mechanism for mitochondrial respiration and oxygen delivery is illustrated in Fig. 5.2. We have confirmed that both the kinetic and thermodynamic control models of mitochondrial respiration are operative in working human muscles [19].

5.3 Principles of Muscle NIRS

Light within the NIR region (700–3,000 nm) shows less scattering and thus better penetration into biological tissue than visible light. However, light absorption by water limits tissue penetration at longer wavelengths (above ~900 nm), leaving the 700–900 nm window for biological monitoring.

The major absorbing compounds in this wavelength region are vascular hemoglobin (Hb), intracellular Mb, and mitochondrial cytochrome c oxidase [10]. Therefore, NIRS measurements rely on O_2-dependent absorption changes that occur in the heme-containing compounds.

The most common commercially available NIRS devices use single-distance continuous-wave light (NIR_{SDCWS}). To calculate changes in oxy-Hb/Mb, deoxy-Hb/Mb, or total-Hb/Mb, the equation of a two- or multiple-wavelength (770–850 nm) method can be applied according to the following Beer-Lambert Law:

$$\Delta \text{OD} = -\log_e(I/I_0) = \varepsilon \text{PL}\, \Delta[C], \quad (5.3)$$

$$\Delta[C] = \Delta \text{OD}/\varepsilon \text{PL}, \quad (5.4)$$

where ε is the extinction coefficient (OD/cm/mM) (= constant), PL is pathlength, $[C]$ is concentration of absorber (mM), I is detected light intensity, I_0 is incident light intensity, and OD is optical density.

A major limitation of NIR_{SDCWS} is that it provides only relative values of tissue oxygenation. The main reason for this lack of quantification by NIR_{SDCWS} is the unknown path of NIR light through biological tissue. The pathlength of NIR light can be measured using other optical approaches, e.g., time-resolved spectroscopy (NIR_{TRS}) [20–23] and phase-modulation spectroscopy (NIR_{PMS}) [23–25]. These approaches provide experimentally observed values of oxygenated and deoxygenated Hb/Mb and Hb/Mb O_2 saturation (SO_2) in skeletal muscle.

With NIR_{TRS}, light is input to tissues in picosecond pulses, and the temporal point spread function (TPSF) is detected as a function of time with picosecond resolution [20, 26]. When a time-correlated single-photon counting system is employed, the distribution of arrival times of a large number of photons follows the TPSF. In principle, the optical properties of tissue can be determined from the TPSF analogous to electronic engineers determining the equivalent circuit of a black box from its response to a voltage or current impulse [26]. Optical properties such as absorption and scattering coefficients can be determined after having an accurate model for light transport in tissues, and represent averaged values for these parameters that best match the monitored data [26].

In NIR_{PMS} instruments, the light source is intensity modulated at radio frequencies (RFs), and measurement is made not only of detected light intensity, but also of its phase shift and modulation depth. Detection of small changes in phase shift and modulation depth is difficult, so almost all these instruments employ some downconversion scheme to bring the RF signal down to an intermediate frequency or audio frequency. In general, two different setups have been used. First, detector gain is modulated at a reference frequency offset by a few kHz from the light source modulation frequency, resulting in direct demodulation within the detector. Second, the detector signal is fed into a double-balanced mixer together with the reference RF, which requires a simpler scheme and much lower RF powers (~10 mW) than the former [26]. The phase shift for typical tissues at frequencies below 200 MHz is linearly related to average optical pathlength. Knowing phase shift enables direct calculation of changes in oxy- and deoxy-hemoglobin concentrations.

Spatially resolved NIR_{SRCWS} (NIR_{SRCWS}) uses multiple light sources coupled to one detector. In NIR_{SRCWS}, the slope of light attenuation versus distance is measured at a distance from the light input, from which the tissue oxygenation index (TOI) can be calculated using photon diffusion theory. Therefore, NIR_{SRCWS} provides relative changes in Hb/Mb and absolute values of SO_2 [27].

With NIR_{SDCWS} measurements, there is an assumption that pathlength does not show significant change during exercise, recovery, and other intervention periods; otherwise, the values obtained are either underestimated or overestimated, as shown in Eqs. 5.3 and 5.4. However, the pathlength of light might vary due to variations in tissue composition, blood volume, and muscle geometry. During and after arterial occlusion, changes in pathlength for forearm muscle ranged from −8.3% to −2.1% at 780 nm and from −2.2% to 0.74% at 830 nm [22]. Changes in pathlength were less than 10% during arterial occlusion with maximum voluntary contraction (MVC) [21]. The change in pathlength at 780 and 830 nm during forearm exercise from moderate to high intensity and hyperemic recovery was small (<8%) relative to the resting level [28]. The differential pathlength factor (DPF) in thigh muscle decreased slightly, but significantly, from baseline (DPF at 690 nm = 5.22, at 830 nm = 4.49, on average) to peak cycle exercise (DPF at 690 nm = 4.88; at 830 nm = 4.27 on average) (−6.5% at 690 nm and −4.9% at 830 nm) [29]. For an accurate evaluation of muscle oxygenation during arterial occlusion, exercise, and recovery, changes in pathlength should be extensively examined in a wide range of exercise mode/intensity.

5.4 Quantification of In-Vivo NIRS Measurements

5.4.1 Mb/Hb

A recent study reported that the spectral peak appearing around 760 nm in deoxygenated muscle tissue included both deoxy-Hb and deoxy-Mb signals, and that the peak position shifted as a linear function of the relative contributions of Hb and Mb to the optical spectra. The investigators reported that Hb accounts for 87% of the overall signal in mouse muscle but for only 20% in human skeletal muscle [30]. ^{1}H-magnetic resonance spectroscopy (^{1}H-MRS) has also been employed to detect the deoxy-Mb signal by measuring the N-delta proton of proximal histidine [31, 32]. Simulation results based on combined measurements of ^{1}H-MRS and NIRS concluded that the overall NIR signal is greater than ~50% Hb [33]. There is also a report indicating differential kinetics between MbO_2 and NIRS measurements. Another study [34] found that, at rest, human tibialis anterior intramuscular O_2 stores (measured by appearance of ^{1}H-NMR deoxy-Mb signal during cuff occlusion) began to decrease after 1 min and that maximal Mb desaturation was achieved at about 6.5 min. Richardson et al. found that deoxy-Mb levels did not appear to increase after 2 min of cuff ischemia [35], while NIRS-measured oxygen signals declined almost immediately and reached near-maximal levels at 4 min [19]. In contrast, It is reported that deoxy-Mb signal correlated with NIR deoxygenation in a study using ^{1}H-MRS and NIRS [36]. These differing conclusions highlight the need for additional study to clarify not only the contribution of Mb to the NIR signal, but also the kinetics and amount of Mb desaturation during exercise under different conditions.

5.4.2 Effect of Multiple Layers

The light path from a light source to the detector follows a banana-shaped curve in which penetration depth into tissue is approximately equal to half the distance between the light source and detector [12]. If light source–detector separation was set at 3 cm, penetration depth was equal to 1–2 cm and measured volume ~4 cm [12, 37]. Hampson et al. studied how the signal from skin might interfere with muscle NIRS measurement, employing NIR_{SDCWS}, changes in skeletal muscle O_2 content kinetics and skin deoxygenation during 10-min ischemia in human forearm [13]. They found that muscle O_2 content began to fall at onset of arterial occlusion. Increased pressure caused skin oxygenation to rapidly fall within 2–3 min, while muscle oxygenation declined within 6 min. Muscle O_2 content was completely depleted (functional anoxia) about 6 min after onset of arterial occlusion. This indicated that the time course of skin deoxygenation changes during ischemia correlate poorly with changes in skeletal muscle. Heating of the thigh at 37°C and 42°C caused a marked increase in cutaneous vascular conductance (CVC) at rest and with exercise, and an increase in SO_2 by several percentage points at rest, but not during exercise. One might thus suppose that skin vasodilatation has a marginal influence on NIRS experiments where skin blood flow can change markedly [38]. A fixed source–detector spacing of 4 or 5 cm ensures more accurate quantitation of the oxygenation changes occurring at the muscular level and minimizes the influence of skin vasculature [39].

Subcutaneous adipose tissue thickness has a substantial influence on signal intensity [40–42]: signal intensity for 5-mm fat thickness mm was reduced by ~0.2 (80% of signal for 0 mm) with a source–detector separation of 30–40 mm, and was further reduced by 0.3–0.6 with a separation of 15–20 mm [41, 43]. Other work documented the influence of adipose tissue thickness ranging from 0 to 15 mm with a source–detector separation of 15–40 mm. The curve was based on the results of Monte Carlo simulation as well as in-vivo experiments [41, 43]:

$$S = \exp\left(-(h/A_1)^2\right) - A_2G(\alpha, \beta), \tag{5.5}$$

where S is normalized measurement sensitivity, h is adipose tissue thickness, $G(\alpha,\beta)$ is a gamma distribution, and constants A_1, A_2, α, and β at a source–detector separation of 15 mm are 6.9, 1.15, 7.86, and 0.80, respectively. Considering that the value of S is mainly determined by h, then corrected values are obtained by dividing the measured values by S.

The influence of adipose tissue thickness in human muscle on NIR spectra was also studied by Monte Carlo simulation in a two-layer model using phantom experiments [44]. The results of this work suggested that subject-to-subject variation in adipose tissue optical coefficients and thickness can be ignored if the thickness is less than 5 mm when the source–detector separation is 40 mm. Subcutaneous adipose tissue thickness should thus be measured for muscle NIRS measurement. In addition, it is preferable to use a 4- to 5-cm source–detector separation for measurements at the muscle level without attenuating signal intensity.

5.5 In-Vitro and In-Vivo Calibration of NIRS Measurements

Several researchers have demonstrated and justified their equations by in-vitro and in-vivo experiments. Chance et al. [12] adopted an "in vitro yeast and blood model" for evaluation of an NIR apparatus. Their model system included intralipid as a scatterer (0.5–1.0%) and blood at concentrations of 10–200 μM hemoglobin. O_2 bubbling caused reoxygenation, while cessation of bubbling caused deoxygenation of hemoglobin due to yeast respiration. Addition of blood correlated with increased in-vivo blood volume. Subtracted signals at 760 and 850 nm showed a linear relationship with hemoglobin deoxygenation [12]. Wilson et al. [45] demonstrated a linear relationship between near-infrared spectroscopic measurements and venous hemoglobin saturation in an animal model. It should be noted that the contribution of Hb/Mb should be different for human and mouse muscle [30]. Mancini et al. found muscle oxygenation and venous O_2 saturation (SvO_2) of human forearm muscles to be closely related during exercise [46]. They also demonstrated that muscle oxygenation decreased with intravascular norepinephrine administration and increased with administration of a vasodilator (nitroprusside). A good association was found between regional quadriceps oxygenation at three different measurement sites and SvO_2 during one-legged dynamic knee extension exercise under normoxic conditions [47]. A good relationship was also found between vastus lateralis oxygenation and femoral arteriovenous O_2 difference (a-vO_2D) during one-legged dynamic knee extension exercise under normoxic as well as hypoxic and hyperoxic conditions [4]. However, several studies have failed to validate NIRS measurement under normoxic conditions

[48, 49]. A possible explanation for this discrepancy is that the NIRS signal contains information on arteriolar, capillary, venular, and intracellular Mb, while the O_2 gradient from arteriole to venule is great during normoxic conditions, such that variations in blood volume from arteriole to venule could alter the NIRS signal without changing venous oxygen signals [40]. Lower oxygen levels during hypoxic conditions would reduce this effect. It is thus broadly accepted that the NIRS oxygenation/deoxygenation signal correlates strongly with changes in SvO_2 and/or a-vO_2D under varying oxygenation status in human muscle.

The use of a physiological calibration in combination with MRS measurements has allowed NIRS to measure muscle oxygen consumption (mVO_2) [19]. The rate of decline of muscle oxygenation during ischemia can be compared with that of muscle PCr in mM/s or a conversion to mVO_2 in mM/s. The transient arterial occlusion method has been used to measure forearm muscle metabolism during exercise [19, 42, 50]. Steady-state exercise at varying intensities was used to provide a range of mVO_2, and resulting mVO_2 values were compared to simultaneous MRS measurements of phosphorus metabolites. A significant correlation was found between mVO_2 and PCr, and ADP concentrations. The linear relationships between exercise intensity and NIRS- and MRS-measured indicators support both thermodynamic [18] and kinetic [15] regulation models of mitochondrial respiration in skeletal muscle. The rate of deoxygenation using transient arterial occlusion has been compared to rate of PCr recovery, a biochemical process of ATP resynthesis via oxidative phosphorylation after exercise (Fig. 5.3). The rate of deoxygenation was quantitatively correlated with the rate of PCr recovery, proving the validity of transient arterial occlusion [50].

Previous studies demonstrated that oxygenation of forearm flexor muscles closely reflects exercise intensity and metabolic rate determined by MRS during exercise and recovery [28, 51], and that muscle oxygenation level (% of arterial occlusion) shows a linear relationship with mVO_2, though in a limited range ($3.2 < mVO_2 < 13.3$-fold of resting), during exercise and recovery [52]. Therefore, where the transient arterial occlusion method is not applicable, muscle oxygenation level can serve to approximate mVO_2.

5.6 Application of NIRS to Physiological Science

Early studies employed onset and recovery kinetics of oxygen saturation to evaluate oxygen use and oxygen delivery [12]. Muscle reoxygenation recovery time reflects the balance of oxygen delivery and oxygen demand in localized muscles. Recovery time measurements are based on extensive study of PCr recovery time, or on the PCr/Pi ratio. Chance et al. [12] compared male elite rowers to female elite rowers and made suggestions for improved performance. They reported a prolonged recovery time, which suggested increased energy deficit when exercise intensity increased. They also compared recovery time in submaximal and maximal work with plasma lactate, and demonstrated a significant correlation between blood lactate and muscle reoxygenation recovery time after exercise. Several studies have reported that muscle reoxygenaiton recovery time after submaximal to maximal exercise is one of the indicators for evaluating muscle oxidative capacity [12, 53]. The deoxy-Hb/Mb pattern during ramp cycling exercise has been monitored to distinguish trained cyclists from physically active subjects [54]. One set of authors proposed a method for noninvasively approximating muscle capillary blood flow kinetics using the kinetics of the primary component of pulmonary O_2 uptake and deoxy-Hb/Mb in humans during exercise [55]. Other investigators compared rate of deoxygenation at onset of intermittent aerobic plantar flexion (7 on and 3 s off, repeatedly), where anaerobic glycolysis is negligible, to muscle oxidative enzyme activity [56], and demonstrated a good correlation between rate of deoxygenation and citrate synthase activity. Thus, we can assume that the rate of onset of deoxygenation reflects muscle oxidative capacity. One might consider that rate of mitochondrial

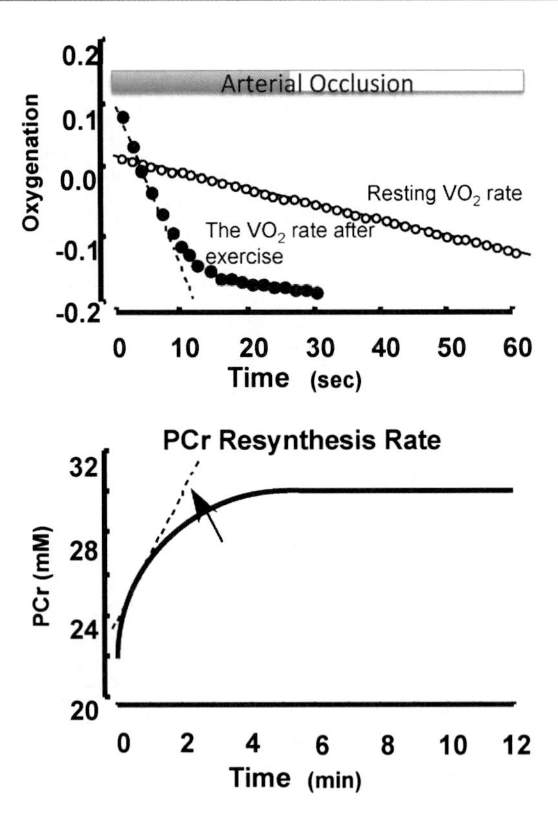

Fig. 5.3 Rate of hemoglobin/myoglobin deoxygenation post-exercise transit ischemia and phosphocreatine (*PCr*) resynthesis rate. Muscle oxygen consumption (VO_2) using the transient arterial occlusion method was compared to rate of PCr recovery, a biochemical process of oxidative ATP resynthesis, at 30 s after exercise. The PCr kinetics is expressed as simulated data (Adapted with permission from J Appl Physiol 90(1):338–344. Copyright © 2001, American Physiological Society)

respiration can be determined by rate of adenine nucleotide translocation and, therefore, that the [ATP]/[ADP] ratio regulates respiratory rate under physiological conditions [16]. However, ATP is maintained constant during submaximal aerobic exercise by creatine kinase equilibrium, so that we need not consider ADP adenine nucleotide translocation during aerobic exercise.

The deoxygenation patterns of the vastus lateralis (VL) and the lateral head of the gastrocnemius (GL) have been examined during graded treadmill exercise [57]. This study found a negative relationship between pulmonary VO_2 and muscle oxygenation level, that the pattern of deoxygenation between VL and GL is somewhat different, and that muscle oxygenation level is associated with pulmonary VO_2. Evaluation of muscle oxygenation in conjunction with systemic oxygen uptake would offer insight into the physiology of healthy and athletic individuals and provide a better exercise prescription for functional improvement. The influence of increased as well as decreased activity on muscle function has also been measured employing NIRS. Most studies have evaluated acute changes in muscle oxygenation during aerobic exercise, but several have also examined high-intensity exercise [58, 59]. In addition, NIRS has been employed to evaluate the training effects of exercise on muscle oxygenation and oxidative metabolism for various types of athletes (e.g., endurance [60, 61] and sprinters [62, 63]) in various sports. Costes et al. [64] attempted to determine whether exercise training–induced adaptations in muscle can be determined by NIRS and found that training does not change the pattern of muscle oxygenation, though a significant relationship was found between blood lactate and muscle oxygenation at the end of exercise. Motobe et al. [65] used NIRS in immobilized forearm muscle to measure changes in skeletal muscle oxidative function and to evaluate the preventive effect of the endurance training protocol on deterioration of skeletal muscle. They found that muscle oxidative function was determined by the time constant for recovery of mVO_2 applying repeated transient arterial occlusions after exercise (see Fig. 5.4). NIRS measurement indicated delayed mVO_2 recovery after exercise during immobilization.

In summary, NIRS can provide useful information on noninvasive monitoring of deconditioning and reconditioning of skeletal muscle oxidative function. However, most studies on the influence of training have been performed using a cross-sectional study design. There is still a need for more longitudinal studies on exercise training that employ NIRS measurements.

Several multichannel NIRS systems have been developed to detect regional differences in muscle oxygenation [66–68]. By simultaneously collecting data from multiple muscle regions, these devices avoid the variability caused by position-dependent differences in muscle oxygenation that plague all single-location measurements. Imaging devices also allow study of regional differences in how skeletal muscle responds to exercise. Another rationale for using multichannel NIRS systems is that measuring at multiple locations provides better agreement between NIRS signal and oxygen saturation in the entire limb [47]. By simultaneously collecting data from multiple muscle regions, these devices avoid the variability caused by position-dependent differences in muscle oxygenation that plague all single-location measurements. Multichannel NIRS also holds an advantage over NMR and PET devices in terms of better time resolution.

5.7 Application of NIRS to Clinical Practice

The strong dependence of skeletal muscle on oxidative metabolism during exercise means that an improvement in the body's oxidative system leads to higher performance during athletic events. On the other hand, impairment of VO_2 and/or DO_2 will limit performance and thus lead to functional deterioration. NIRS is suitable for measuring attenuation of VO_2 and DO_2 in patients with various diseases and possesses great advantages in terms of portability and real-time monitoring over conventional technologies.

5.7.1 Congestive Heart Failure (CHF)

NIRS has also been employed to evaluate skeletal muscle oxygenation in patients with heart disease. Patients with congestive heart failure (CHF) [45, 69–71] showed greater deoxygenation compared with controls partly due to heart pump failure and the consequent skeletal muscle hypoperfusion. The authors of [71] found a correlation between changes in total hemoglobin (tHb) and leg vessel conductance in patients with and without cardiac dysfunction during submaximal dynamic exercise. This result indicated that tHb reflects muscle vasodilation in these patients and that tHb can be used to assess vascular conductance in this patient group. Another study [72] demonstrated that changes in deoxygenation during submaximal exercise are steeper but peak deoxygenation lower in heart transplant recipients (HTRs) than in control subjects. The authors suggested that NIRS allows for detection of impairment of both DO_2 and O_2 extraction in HTR skeletal muscle. NIRS has also been used to examine the effect of pharmacological treatment on blunted microvascular oxygen delivery to muscle [73] and the effect of rehabilitation on muscle oxygenation [74] in patients with CHF.

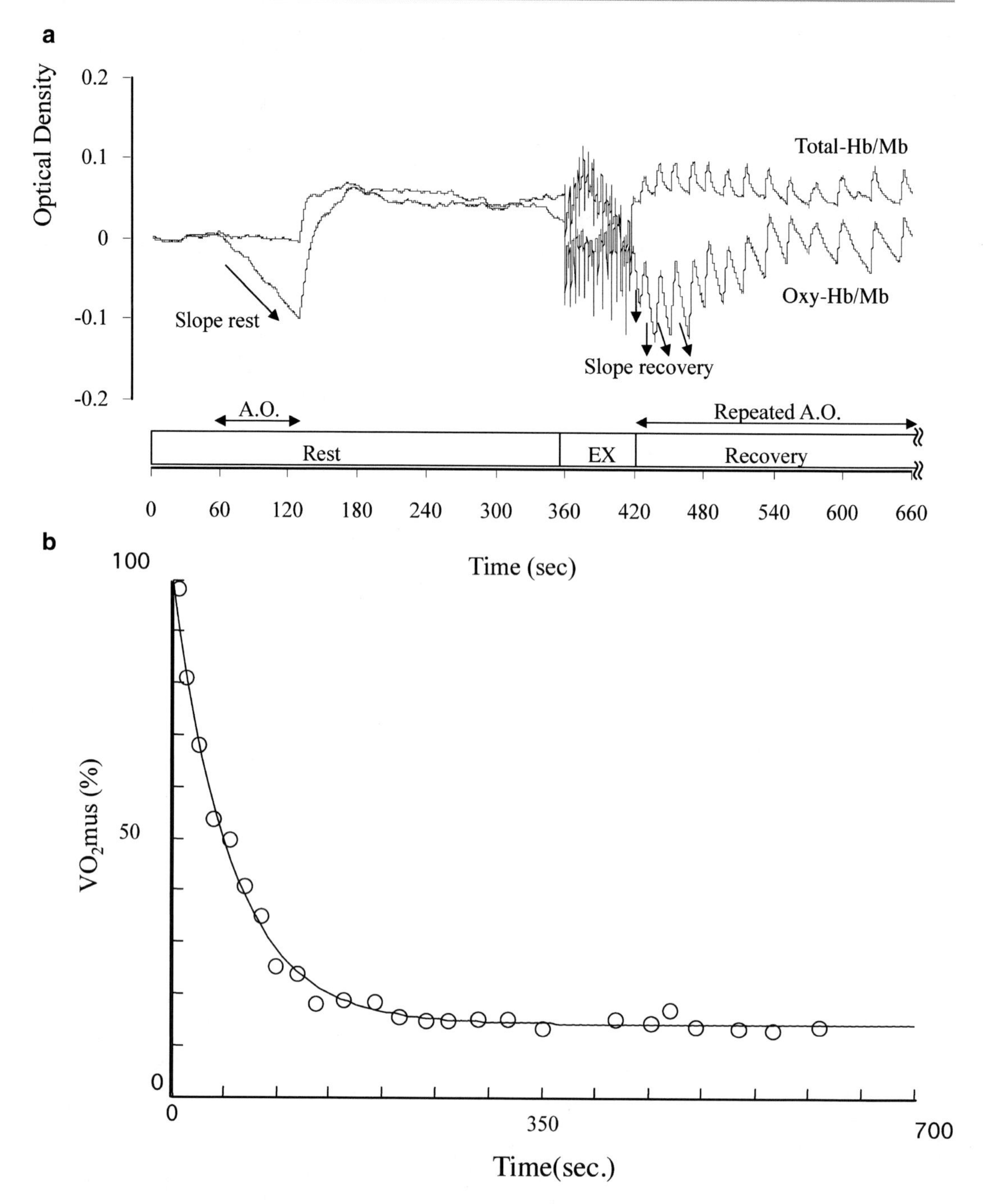

Fig. 5.4 (a) **Schematic representation of muscle oxygen consumption (VO_{2mus}mus) and typical changes in muscle oxygenated Hb/Mb**. Schematic representation of VO_2mus and typical changes in muscle oxygenated Hb/Mb at rest and during exercise and recovery. VO_2mus was calculated from the rate of decline of oxygenated Hb/Mb during arterial occlusion (A.O.) at rest (Slope rest), and recovery (Slope recovery). (**b**) **Typical kinetics of VO_2mus recovery after exercise**. VO_2mus during recovery was compared with that at rest, normalized by the maximum value, and plotted as a function of time in seconds. Time constant of VO_2mus for this subject was 56 s (Adapted with permission from Dyn Med 3 Galadriel 3:2. Copyright 2004, BioMed Central)

5.7.2 Patients with Chronic Obstructive Pulmonary Disease (COPD)

Patients with COPD frequently develop skeletal muscle and vascular abnormalities as complications of their disease, similar to those with heart disease [75, 76]. Muscle reoxygenation recovery after exercise has been shown to correlate with pulmonary VO_2 recovery kinetics in patients with COPD [77]. NIRS has also been employed to obtain the time constant of reoxygenation recovery during three work exercise tests, one below and two above the lactic acidosis threshold [78], where patients trained for 6 weeks on a cycle ergometer at high work rates. These investigators found significant correlations between changes in oxidative enzyme activity and changes in recovery time constant and endurance. Leg training is known to accelerate speed of reoxygenation of the vastus lateralis muscle after exercise, and this improvement has been correlated with changes in oxidative enzymes [79].

5.7.3 Neuromuscular Disorders

NIRS has also been employed to study patients with neuromuscular disorders. An increase in muscle oxygenation (paradoxical oxygenation) at onset of treadmill exercise was been reported in patients with cytochrome c oxidase deficiency [80], metabolic myopathy [81], and Friedreich's ataxia [82]. This paradoxical oxygenation, a diagnostic indicator, is due to a combination of impaired mVO_2 and a normal physiological increase of DO_2 (vasodilatation) stimulated by muscle pump and/or myogenic activity. Grassi et al. [83] tested patients with mitochondrial myopathy (MM) or myophosphorylase deficiency (McArdle's disease, McA) for changes in capacity for O_2 extraction, maximal aerobic power, and exercise tolerance during cycle exercise using NIRS. They found that peak deoxygenation during exercise, an index of O_2 extraction, was lower in MM and McA than in controls.

5.7.4 Peripheral Arterial Disease (PAD)

Several studies have used rate of recovery of reoxygenation to evaluate DO_2 in the calf muscles of patients with peripheral artery disease (PAD) [84]. In a rather extensive study, the authors of [85] reported good agreement between faster PCr recovery kinetics and better oxygenation kinetics measured by NIRS [85]. A slower rate of calf reoxygenation after exercise is a consistent finding in PAD patients (see [86–88]).

Patients with varying severity of PAD have been classified by using patterns of calf oxygenation kinetics during treadmill exercise and recovery [87]. Impaired muscle O_2 usage at rest [89] and at onset of exercise [90] have been observed as well. The authors of [84] found a good correlation between Doppler pressure waveform measurements and ankle arm systolic pressures (AAI) and the NIRS recovery time constant [84]. Mohler et al. [86] examined the interaction between PAD and presence or absence of diabetes mellitus (DM) employing changes in muscle capillary blood expansion and reoxygenation recovery, and found that capillary blood expansion is reduced in patients with DM regardless of existence of PAD. This parameter might thus be a good indicator of vascular impairment in patients with DM. Peripheral venous occlusive disease (PVOD) has also been tested using NIRS [91–93], and investigators were able to distinguish between successfully treated patients and those with deep vein thrombosis after 12 months. So we can conclude that NIRS is able to identify and quantify severity of DO_2 in patients with PAD and PVOD.

5.7.5 Spinal Cord Injury (SCI)

NIRS has also been employed to evaluate changes in oxygenation that occur in the leg muscles of patients with spinal cord injury (SCI). Bhambhani et al. [94] found lower muscle deoxygenation during maximal exercise and faster changes in muscle deoxygenation with respect to pVO_2 during functional electrical stimulation cycle exercise in SCI patients compared to healthy subjects. NIRS has been used to evaluate potential SCI therapies. The authors of [95] placed motor-complete SCI subjects and neurologically normal controls on a gait-training apparatus that enabled subjects to stand and move their legs passively, and found that oxygenation gradually increased and deoxygenation decreased in the experimental group, quite different from the response in normal controls. This result demonstrated that passive leg movement can not only induce muscular activity but can also alter muscle oxygenation in a paralyzed lower leg, and that induced muscular activity seems to correlate with increased muscle perfusion. In another NIRS study [96], SCI patients underwent electrical stimulation training (45 min daily, 3 days a week for 10 weeks) with different muscle oxygenation loads in paralyzed lower limbs. The vastus lateralis muscle of statically trained legs was found to show significantly increased oxidative capacity compared to both baseline and dynamically trained legs. This study demonstrated that NIRS is able to detect attenuated muscle deoxygenation after static training.

5.7.6 End-Stage Renal Disease (ESRD)

Muscle oxygenation and metabolism were examined using NIRS in children with end-stage renal disease (ESRD) before and after kidney transplantation and compared to controls during submaximal handgrip exercise [97]. These investigators looked at rate of deoxygenation, an indicator of mVO_2, during transient arterial occlusion post-exercise and recovery time to reoxygenation after exercise. Both parameters were significantly improved after renal transplantation but not significantly different from controls (Fig. 5.5). Another group [98] employed NIRS to evaluate the potential for vascular and metabolic dysfunction in patients with renal failure, the effect of handgrip exercise training on forearm vasodilator responses, and forearm vasodilator responses to 3-min arterial occlusion in patients receiving hemodialysis, but found no improvement in vasodilator response after exercise training, and vasodilator response estimated by maximum oxygenation after release of arterial occlusion was significantly smaller in renal failure patients compared to controls. These studies show that NIRS can be used to detect muscle hypoperfusion in patients with renal failure and to find the functional alterations of muscle oxidative metabolism that occur after renal transplantation.

The results of clinical NIRS application suggest that it is a promising tool for noninvasively monitoring metabolic impairment in a follow-up setting, and in assessment of therapies and interventions.

5.8 Summary

There are several issues involved with in-vivo muscle NIRS measurement: (a) the origin of the NIR signal (from arterioles, capillary, and venules, as well as from Hb and Mb); (b) the NIR penetration depth or measurement area in tissue with varying source–detector arrangements in a multilayer model, including the effect of non-muscular tissue; (c) changes in optical properties within a wide

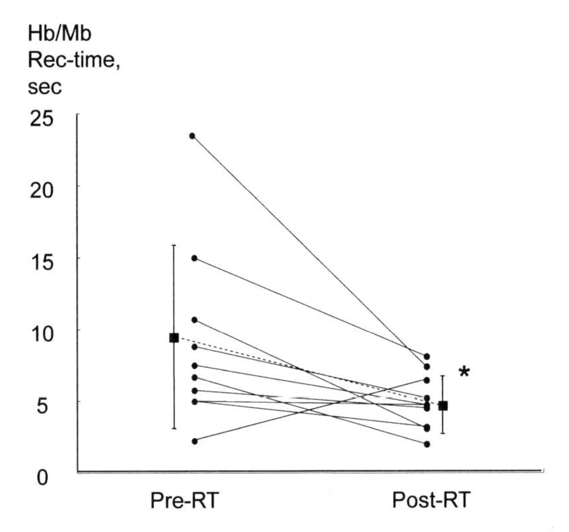

Fig. 5.5 Changes in recovery time (Rec-time) for Hb/Mb reoxygenation after renal transplantation (RT) in 10 renal transplant recipients. Individual data shown in *black circles* and *solid lines*; mean and SD shown in *black squares* and *dotted line*. Significant difference $p < 0.05$. Recovery time after transplantation was shortened in nine patients and delayed in one patient, who showed the shortest TR before RT (Adapted with permission from Am J Kidney Dis 48:473–480. Copyright © 2006, Elsevier)

range of tissue oxygenation status; (d) the wide variety of subjects; and (e) the choice of exercise modality. Nevertheless, there is little question that NIRS is suitable and quite useful for measuring changes in mVO_2 and DO_2 in healthy subjects as well as in patients with various disease states (e.g., chronic heart failure, chronic obstructive pulmonary disease, various muscle diseases, peripheral vascular disease, spinal cord injury, renal failure). In addition, NIRS has great advantages over conventional technologies in terms of portability and real-time monitoring.

Acknowledgments I would like to acknowledge funding support from the Ministry of Education, Culture, Sports, Science, and Technology of Japan.

Problems

5.1 How would you quantify muscle NIR signals?
5.2 List the various muscle NIR indicators. Which indicator reflects muscle oxidative function? How?

Further Reading

Ferrari M, Muthalib M, Quaresima V (2011) The use of near-infrared spectroscopy in understanding skeletal muscle physiology: recent developments. Philos Trans A Math Phys Eng Sci 369:4577–4590.

Hamaoka T, McCully K, Quaresima V, Yamamoto K, Chance B (2007) Near-infrared spectroscopy/imaging for monitoring muscle oxygenation and oxidative metabolism in healthy and diseased humans. J Biomed Opt 12(6):62105–62120.

Hamaoka T, McCully K, Quaresima V, Yamamoto K, Chance B (2011) The use of muscle near-infrared spectroscopy in sport, health, and medical sciences: recent developments. Philos Trans A Math Phys Eng Sci 369:4591–4604.
Yamamoto K, Niwayama M, Lin L, Shiga T, Kudo N, Takahashi M (1998) Accurate NIRS measurement of muscle oxygenation by correcting the influence of a subcutaneous fat layer. Proc SPIE 3194:166–173.

References

1. Welch HG, Bonde-Petersen F, Graham T, Klausen K, Secher N (1977) Effects of hyperoxia on leg blood flow and metabolism during exercise. J Appl Physiol 42:385–390
2. Gayeski TE, Honig CR (1983) Direct measurement of intracellular O_2 gradients; role of convection and myoglobin. Adv Exp Med Biol 159:613–621
3. Bhambhani YN (2004) Muscle oxygenation trends during dynamic exercise measured by near infrared spectroscopy. Can J Appl Physiol 29:504–523
4. Boushel B, Langberg H, Olesen J, Gonzales-Alonzo J, Bulow J, Kjaer M (2001) Monitoring tissue oxygen availability with near infrared spectroscopy (NIRS) in health and disease. Scand J Med Sci Sports 11:213–222
5. Ferrari M, Mottola L, Quaresima V (2004) Principles, techniques, and limitations of near infrared spectroscopy. Can J Appl Physiol 29:463–487
6. Hamaoka T, McCully K, Quaresima V, Yamamoto K, Chance B (2007) Near-infrared spectroscopy/imaging for monitoring muscle oxygenation and oxidative metabolism in healthy and diseased humans. J Biomed Opt 12(6):62105–62120
7. Millikan GA (1933) A simple photoelectric colorimeter. J Physiol 79:152–157
8. Chance B (1954) Spectrophotometry of intracellular respiratory pigments. Science 120:767–775
9. Chance B, Connelly CM (1957) A method for the estimation of the increase in concentration of adenosine diphosphate in muscle sarcosomes following a contraction. Nature 179:1235–1237
10. Jöbsis FF (1977) Noninvasive, infrared monitoring of cerebral and myocardial oxygen sufficiency and circulatory parameters. Science 198:1264–1267
11. McCully KK, Kakihira H, Vandenborne K, Kent-Braun J (1991) Noninvasive measurements of activity-induced changes in muscle metabolism. J Biomech 24:153–161
12. Chance B, Dait MT, Zhang C, Hamaoka T, Hagerman F (1992) Recovery from exercise-induced desaturation in the quadriceps muscles of elite competitive rowers. Am J Physiol 262:C766–C775
13. Hampson NB, Piantadosi CA (1988) Near infrared monitoring of human skeletal muscle oxygenation during forearm ischemia. J Appl Physiol 64(6):2449–2457
14. Chance B, Williams GR (1955) Respiratory enzymes in oxidative phosphorylation, I: kinetics of oxygen utilization. J Biol Chem 217:409–427
15. Chance B, Leigh JS, Kent-Braun JA, McCully K, Nioka S, Clark BJ, Maris JM, Graham T (1986) Multiple controls of oxidative metabolism in living tissues as studied by phosphorus magnetic resonance. Proc Natl Acad Sci USA 83:9458–9462
16. Klingenberg M (1980) The ADP–ATP translocation in mitochondria, a membrane potential controlled transport. J Membr Biol 56(2):97–105
17. Holian A, Owen CS, Wilson DF (1977) Control of respiration in isolated mitochondria: quantitative evaluation of the dependence of respiratory rates on [ATP], [ADP], and [Pi]. Arch Biochem Biophys 181:164–171
18. Meyer RA (1988) A linear model of muscle respiration explains monoexponential phosphocreatine changes. Am J Physiol 254:C548–C553
19. Hamaoka T, Iwane H, Shimomitsu T, Katsumura T, Murase N, Nishio S, Osada T, Kurosawa Y, Chance B (1996) Noninvasive measures of oxidative metabolism on working human muscles by near-infrared spectroscopy. J Appl Physiol 81:1410–1417
20. Chance B, Nioka S, Kent-Braun J, McCully K, Fountain M, Greenfeld R, Holtom G (1988) Time-resolved spectroscopy of hemoglobin and myoglobin in resting and ischemic muscle. Anal Biochem 174:698–707
21. Ferrari M, Wei Q, Carraresi L, De Blasi RA, Zaccanti G (1992) Time-resolved spectroscopy of the human forearm. J Photochem Photobiol B 16:141–153
22. Hamaoka T, Katsumura T, Murase N, Nishio S, Osada T, Sako T, Higuchi H, Kurosawa Y, Shimomitsu T, Miwa M, Chance B (2000) Quantification of ischemic muscle deoxygenation by near infrared time-resolved spectroscopy. J Biomed Opt 5:102–105
23. Delpy DT, Cope M, van der Zee P, Arridge S, Wray S, Wyatt J (1988) Estimation of optical pathlength through tissue from direct time of flight measurement. Phys Med Biol 33:1433–1442
24. Wolf M, Wolf U, Choi JH, Gupta R, Safonova LP, Paunescu LA, Michalos A, Gratton E (2002) Functional frequency-domain near-infrared spectroscopy detects fast neuronal signal in the motor cortex. Neuroimage 17:1868–1875

25. Duncan A, Meek JH, Clemence M, Elwell CE, Tyszczuk L, Cope M, Delpy DT (1995) Optical pathlength measurements on adult head, calf and forearm and the head of the newborn infant using phase resolved optical spectroscopy. Phys Med Biol 40:295–304
26. Delpy DT, Cope M (1997) Quantification in tissue near-infrared spectroscopy. Philos Trans Roy Soc Lond B 352:649–659
27. Quaresima V, Homma S, Azuma K, Shimizu S, Chiarotti F, Ferrari M, Kagaya A (2001) Calf and shin muscle oxygenation patterns and femoral artery blood flow during dynamic plantar flexion exercise in humans. Eur J Appl Physiol Occup Physiol 84:387–394
28. Hamaoka T, Osada T, Murase N, Sako T, Higuchi H, Kurosawa Y, Miwa M, Katsumura T, Chance B (2003) Quantitative evaluation of oxygenation and metabolism in the human skeletal muscle. Opt Rev 10(5):493–497
29. Ferreira LF, Hueber DM, Barstow TJ (2007) Effects of assuming constant optical scattering on measurements of muscle oxygenation by near-infrared spectroscopy during exercise. J Appl Physiol 102:358–367
30. Marcinek DJ, Amara CE, Matz K, Conley KE, Schenkman KA (2007) Wavelength shift analysis: a simple method to determine the contribution of hemoglobin and myoglobin to in vivo optical spectra. Appl Spectrosc 61(6):665–669
31. Wang ZY, Noyszewski EA, Leigh JS Jr (1990) In vivo MRS measurement of deoxymyoglobin in human forearms. Magn Reson Med 14:562–567
32. Mole PA, Chung Y, Tran TK, Sailasuta N, Hurd R, Jue T (1999) Myoglobin desaturation with exercise intensity in human gastrocnemius muscle. Am J Physiol 277:R173–R180
33. Nioka S, Wang DJ, Im J, Hamaoka T, Wang ZJ, Leigh JS, Chance B (2006) Simulation of Mb/Hb in NIRS and oxygen gradient in the human and canine skeletal muscles using H-NMR and NIRS. Adv Exp Med Biol 578:223–228
34. Lanza IR, Tevald MA, Befroy DE, Kent-Braun JA (2010) Intracellular energetics and critical PO_2 in resting ischemic human skeletal muscle in vivo. Am J Physiol Regul Integr Comp Physiol 299:R1415–R1422
35. Richardson RS, Newcomer SC, Noyszewski EA (2001) Skeletal muscle intracellular PO_2 assessed by myoglobin desaturation: response to graded exercise. J Appl Physiol 91:2679–2685
36. Tran TK, Sailasuta N, Kreutzer U, Hurd R, Chung Y, Mole P, Kuno S, Jue T (1999) Comparative analysis of NMR and NIRS measurements of intracellular PO_2 in human skeletal muscle. Am J Physiol 276:R1682–R1690
37. Chance B, Nioka S, Kent J, McCully K, Fountain M, Greenfeld R, Holtom G (1988) Time-resolved spectroscopy of hemoglobin and myoglobin in resting and ischemic muscle. Anal Biochem 174:698–707
38. Tew GA, Ruddock AD, Saxton JM (2010) Skin blood flow differentially affects near-infrared spectroscopy-derived measures of muscle oxygen saturation and blood volume at rest and during dynamic leg exercise. Eur J Appl Physiol 110(5):1083–1089
39. Ferrari M, Cettolo V, Quaresima V (2006) Light source-detector spacing of near-infrared-based tissue oximeters and the influence of skin blood flow. J Appl Physiol 100(4):1426–1427
40. McCully KK, Hamaoka T (1998) Near-infrared spectroscopy: what can it tell us about oxygen saturation in skeletal muscle? Exerc Sport Sci Rev 28:123–127
41. Yamamoto K, Niwayama M, Lin L, Shiga T, Kudo N, Takahashi M (1998) Accurate NIRS measurement of muscle oxygenation by correcting the influence of a subcutaneous fat layer. Proc SPIE 3194:166–173
42. Van Beekvelt MC, Borghuis MS, Van Engelen BG, Wevers RA, Colier WN (2001) Adipose tissue thickness affects in vivo quantitative near-IR spectroscopy in human skeletal muscle. Clin Sci (Lond) 101:21–28
43. Niwayama M, Yamamoto K, Kohata D, Hirai K, Kudo N, Hamaoka T, Kime R, Katsumura T (2002) A 200-channel imaging system of muscle oxygenation using CW near-infrared spectroscopy. IEICE Trans Inf Syst E85-D(1): 115–123
44. Yang Y, Soyemi OO, Landry MR, Soller BR (2005) Influence of a fat layer on the near infrared spectra of human muscle: quantitative analysis based on two-layered Monte Carlo simulations and phantom experiments. Opt Express 13:1570–1579
45. Wilson JR, Mancini DM, McCully K, Ferraro N, Lanoce V, Chance B (1989) Noninvasive detection of skeletal muscle underperfusion with near-infrared spectroscopy in patients with heart failure. Circulation 80:1668–1674
46. Mancini DM, Bolinger L, Li H, Kendrick K, Chance B, Wilson JR (1994) Validation of near-infrared spectroscopy in humans. J Appl Physiol 77:2740–2747
47. Esaki K, Hamaoka T, Radegran G, Boushel R, Hansen J, Katsumura T, Haga S, Mizuno M (2005) Association between regional quadriceps oxygenation and blood oxygen saturation during normoxic one-legged dynamic knee extension. Eur J Appl Physiol Occup Physiol 95:361–370
48. Costes F, Barthelemy JC, Feasson L, Busso T, Geyssant A, Denis C (1996) Comparison of muscle near-infrared spectroscopy and femoral blood gases during steady-state exercise in humans. J Appl Physiol 80:1345–1350
49. MacDonald MJ, Tarnopolsky MA, Green HJ, Hughson RL (1999) Comparison of femoral blood gases and muscle near-infrared spectroscopy at exercise onset in humans. J Appl Physiol 86:687–693

50. Sako T, Hamaoka T, Higuchi H, Kurosawa Y, Katsumura T (2001) Validity of NIR spectroscopy for quantitatively measuring muscle oxidative metabolic rate in exercise. J Appl Physiol 90(1):338–344
51. Boushel R, Pott F, Madsen P, Radegran G, Nowak M, Quistorff B, Secher N (1998) Muscle metabolism from near infrared spectroscopy during rhythmic handgrip in humans. Eur J Appl Physiol Occup Physiol 79:41–48
52. Homma T, Hamaoka T, Sako T, Murakami M, Esaki K, Kime R, Ueda C, Nagasawa T, Katsumura T (2005) Muscle oxidative metabolism accelerates with mild acidosis during incremental intermittent isometric plantar flexion exercise. Dyn Med 4:2
53. Ichimura S, Murase N, Osada T, Kime R, Homma T, Ueda C, Nagasawa T, Motobe M, Hamaoka T, Katsumura T (2006) Age and activity status affect muscle reoxygenation time after maximal cycling exercise. Med Sci Sports Exerc 38:1277–1281
54. Boone J, Koppo K, Barstow TJ, Bouckaert J (2009) Pattern of deoxy[Hb + Mb] during ramp cycle exercise: influence of aerobic fitness status. Eur J Appl Physiol 105(6):851–859
55. Ferreira LF, Townsend DK, Lutjemeier BJ, Barstow TJ (2005) Muscle capillary blood flow kinetics estimated from pulmonary O_2 uptake and near-infrared spectroscopy. J Appl Physiol 98(5):1820–1828
56. Hamaoka T, Mizuno M, Katsumura T, Osada T, Shimomitsu T, Quistorff B (1998) Correlation between indicators determined by near infrared spectroscopy and muscle fiber types in humans. Jpn J Appl Physiol 28(5):339–344
57. Hiroyuki H, Hamaoka T, Sako T, Nishio S, Kime R, Murakami M, Katsumura T (2002) Oxygenation in vastus lateralis and lateral head of gastrocnemius during treadmill walking and running in humans. Eur J Appl Physiol 87:343–349
58. Bae S, Hamaoka T, Katsumura T, Shiga T, Ohno H, Haga S (2000) Comparison of muscle oxygen consumption measured by near infrared continuous wave spectros copy during supramaximal and intermittent pedalling exercise. Int J Sports Med 21:168–174
59. Bhambhani Y, Maikala R, Esmail S (2001) Oxygenation trends in vastus lateralis muscle during incremental and intense anaerobic cycle exercise in young men and women. Eur J Appl Physiol 84:547–556
60. Neary JP (2004) Application of near infrared spectroscopy to exercise sports science. Can J Appl Physiol 29:488–503
61. Legrand R, Ahmaidi S, Moalla W, Chocquet D, Marles A, Prieur F, Mucci P (2005) O_2 arterial desaturation in endurance athletes increases muscle deoxygenation. Med Sci Sports Exerc 37:782–788
62. Ding H, Wang G, Lei W, Wang R, Huang L, Xia Q, Wu J (2001) Non-invasive quantitative assessment of oxidative metabolism in quadriceps muscles by near infrared spectroscopy. Br J Sports Med 35:441–444
63. Hoffman JR, Im J, Rundell KW, Kang J, Nioka S, Spiering BA, Kime R, Chance B (2003) Effect of muscle oxygenation during resistance exercise on anabolic hormone response. Med Sci Sports Exerc 35:1929–1934
64. Costes F, Prieur F, Feasson L, Geyssant A, Barthelemy JC, Denis C (2001) Influence of training on NIRS muscle oxygen saturation during submaximal exercise. Med Sci Sports Exerc 33:1484–1489
65. Motobe M, Murase N, Osada T, Homma T, Ueda C, Nagasawa T, Kitahara A, Ichimura S, Kurosawa Y, Katsumura T, Hoshika A, Hamaoka T (2004) Noninvasive monitoring of deterioration in skeletal muscle function with forearm cast immobilization and the prevention of deterioration. Dyn Med 3:2
66. Kime R, Im J, Moser D, Lin Y, Nioka S, Katsumura T, Chance B (2005) Reduced heterogeneity of muscle deoxygenation during heavy bicycle exercise. Med Sci Sports Exerc 37:412–417
67. Miura H, McCully K, Hong L, Nioka S, Chance B (2001) Regional difference of muscle oxygen saturation and blood volume during exercise determined by near infrared imaging device. Jpn J Physiol 51:599–606
68. Wolf U, Wolf M, Choi JH, Paunescu LA, Safonova LP, Michalos A, Gratton E (2003) Mapping of hemodynamics on the human calf with near infrared spectroscopy and the influence of the adipose tissue thickness. Adv Exp Med Biol 510:225–230
69. Hanada A, Okita K, Yonezawa K, Ohtsubo M, Kohya T, Murakami T, Nishijima H, Tamura M, Kitabatake A (2000) Dissociation between muscle metabolism and oxygen kinetics during recovery from exercise in patients with chronic heart failure. Heart 83:161–166
70. Matsui S, Bolinger L, Li H, Kendrick K, Chance B, Wilson JR (1995) Assessment of working muscle oxygenation in patients with chronic heart failure. Am Heart J 125:690–695
71. Watanabe S, Ishii C, Takeyasu N, Ajisaka R, Nishina H, Morimoto T, Sakamoto K, Eda K, Ishiyama M, Saito T, Aihara H, Arai E, Toyama M, Shintomi Y, Yamaguchi I (2005) Assessing muscle vasodilation using near-infrared spectroscopy in cardiac patients. Circ J 69:802–814
72. Lanfranconi F, Borrelli E, Ferri A, Porcelli S, Maccherini M, Chiavarelli M, Grassi B (2006) Noninvasive evaluation of skeletal muscle oxidative metabolism after heart transplant. Med Sci Sports Exerc 38:1374–1383
73. Sperandio PA, Borghi-Silva A, Barroco A, Neder JA (2009) Microvascular oxygen delivery-to-utilization mismatch at the onset of heavy-intensity exercise in optimally treated patients with CHF. Am J Physiol Heart Circ Physiol 297:H1720–H1728

74. Gerovasili V, Drakos S, Kravari M, Malliaras K, Karatzanos E, Dimopoulos S, Tasoulis A, Anastasiou-Nana M, Roussos C, Nanas S (2009) Physical exercise improves the peripheral microcirculation of patients with chronic heart failure. J Cardiopulm Rehabil Prev 29:385–391
75. Serres I, Hayot M, Prefaut C, Mercier J (1998) Skeletal muscle abnormalities in patients with COPD: contribution to exercise intolerance. Med Sci Sports Exerc 30:1019–1027
76. Maltais F, LeBlanc P, Jobin J, Casaburi R (2000) Peripheral muscle dysfunction in chronic obstructive pulmonary disease. Clin Chest Med 21:665–677
77. Okamoto T, Kanazawa H, Hirata K, Yoshikawa J (2003) Evaluation of oxygen uptake kinetics and oxygen kinetics of peripheral skeletal muscle during recovery from exercise in patients with chronic obstructive pulmonary disease. Clin Physiol Funct Imaging 23:257–262
78. Puente-Maestu L, Tena T, Trascasa C, Perez-Parra J, Godoy R, Garcia MJ, Stringer WW (2003) Training improves muscle oxidative capacity and oxygenation recovery kinetics in patients with chronic obstructive pulmonary disease. Eur J Appl Physiol Occup Physiol 88:580–587
79. Rondelli RR, Dal Corso S, Simões A, Malaguti C (2009) Methods for the assessment of peripheral muscle fatigue and its energy and metabolic determinants in COPD. J Bras Pneumol 35:1125–1135
80. Bank W, Chance B (1994) An oxidative defect in metabolic myopathies: diagnosis by noninvasive tissue oximetry. Ann Neurol 36:830–837
81. Grassi B, Porcelli S, Marzorati M, Lanfranconi F, Vago P, Marconi C, Morandi L (2009) Metabolic myopathies: functional evaluation by analysis of oxygen uptake kinetics. Med Sci Sports Exerc 41:2120–2127
82. Lynch DR, Lech G, Farmer JM, Balcer LJ, Bank W, Chance B, Wilson RB (2002) Near infrared muscle spectroscopy in patients with Friedreich's ataxia. Muscle Nerve 25:664–673
83. Grassi B, Marzorati M, Lanfranconi F, Ferri A, Longaretti M, Stucchi A, Vago P, Marconi C, Morandi L (2007) Impaired oxygen extraction in metabolic myopathies: detection and quantification by near-infrared spectroscopy. Muscle Nerve 35:510–520
84. McCully KK, Halber C, Posner JD (1994) Exercise-induced changes in oxygen saturation in the calf muscles of elderly subjects with peripheral vascular disease. J Gerontol 49:B128–B134
85. McCully KK, Iotti S, Kendrick K, Wang Z, Posner JD, Leigh J Jr, Chance B (1994) Simultaneous in vivo measurements of HbO_2 saturation and PCr kinetics after exercise in normal humans. J Appl Physiol 77:5–10
86. Mohler ER 3rd, Lech G, Supple GE, Wangb H, Chance B (2006) Impaired exercise-induced blood volume in type 2 diabetes with or without peripheral arterial disease measured by continuous-wave near-infrared spectroscopy. Diabetes Care 29:1856–1859
87. Komiyama T, Shigematsu H, Yasuhara H, Hosoi Y, Muto T (1996) An objective evaluation of muscle oxygen content in claudicants receiving drug therapy. Int Angiol 15:215–218
88. Komiyama T, Shigematsu H, Yasuhara H, Muto T (2000) Near-infrared spectroscopy grades the severity of intermittent claudication in diabetics more accurately than ankle pressure measurement. Br J Surg 87:459–466
89. Malagoni AM, Felisatti M, Mandini S, Mascoli F, Manfredini R, Basaglia N, Zamboni P, Manfredini F (2010) Resting muscle oxygen consumption by near-infrared spectroscopy in peripheral arterial disease: a parameter to be considered in a clinical setting? Angiology 61:530–536
90. Bauer TA, Brass EP, Hiatt WR (2004) Impaired muscle oxygen use at onset of exercise in peripheral arterial disease. J Vasc Surg 40:488–493
91. Pedersen BL, Baekgaard N, Quistorff B (2009) Muscle mitochondrial function in patients with type 2 diabetes mellitus and peripheral arterial disease: implications in vascular surgery. Eur J Vasc Endovasc Surg 38:356–364
92. Hosoi Y, Yasuhara H, Shigematsu H, Aramoto H, Komiyama T, Muto T (1997) A new method for the assessment of venous insufficiency in primary varicose veins using near-infrared spectroscopy. J Vasc Surg 26:53–60
93. Yamaki T, Nozaki M, Sakurai H, Takeuchi M, Soejima K, Kono T (2006) The utility of quantitative calf muscle near-infrared spectroscopy in the follow-up of acute deep vein thrombosis. J Thromb Haemost 4:800–806
94. Bhambhani Y, Tuchak C, Burnham R, Jeon J, Maikala R (2000) Quadriceps muscle deoxygenation during functional electrical stimulation in adults with spinal cord injury. Spinal Cord 38:630–638
95. Kawashima N, Nakazawa K, Akai M (2005) Muscle oxygenation of the paralyzed lower limb in spinal cord-injured persons. Med Sci Sports Exerc 37:915–921
96. Crameri RM, Cooper P, Sinclair PJ, Bryant G, Weston A (2004) Effect of load during electrical stimulation training in spinal cord injury. Muscle Nerve 29:104–111
97. Matsumoto N, Ichimura S, Hamaoka T, Osada T, Hattori M, Miyakawa S (2006) Impaired muscle oxygen metabolism in uremic children: improved after renal transplantation. Am J Kidney Dis 48:473–480
98. Kuge N, Suzuki T, Isoyama S (2005) Does handgrip exercise training increase forearm ischemic vasodilator responses in patients receiving hemodialysis? Tohoku J Exp Med 207:303–312

Intracellular Oxygen Dynamics Observed by NIRS During Skeletal Muscle Contraction

6

Kazumi Masuda

6.1 Introduction

Cellular respiration depends upon a coordinated response of the cardiovasculature and metabolism to meet changing energy demands in muscle. Even though adjustments in blood flow, O_2 gradient, and myoglobin (Mb) saturation will enhance O_2 flux to the mitochondria at initiation of contraction, the relative contribution of each component remains an issue for contentious debate [1]. In particular, the roles of Mb and Hb have stirred much controversy.

Mb is the first protein whose three-dimensional structure was determined by x-ray analysis (http://nobelprize.org/nobel_prizes/chemistry/laureates/1962/kendrew-lecture.pdf). It is a typical globular protein with heme, being found in most mammalian muscle cells. Many studies have endeavored to discover its ex-vivo and in-vivo functions, and some have suggested that it stores oxygen and facilitates oxygen diffusion within muscle cells. The former is particularly important for such diving animals as whales, seals, and penguins, while the latter remains controversial with respect to in-vivo skeletal muscles. In contrast, some researchers have ascribed no significant role for Mb in supplying O_2 during muscle contraction. Since it has an extremely high affinity for O_2, it cannot readily release its O_2 store. As a results, Hb must then supply all the O_2 from the onset of contraction [2–5]. This viewpoint is the underpinning for interpretation of the noninvasive near-infrared spectroscopy (NIRS) data [6, 7]. Although NIRS cannot be used to discriminate between Hb and Mb signals, many researchers have assumed that NIRS can be employed to monitor Hb oxygen saturation and desaturation kinetics. As a consequence, NIRS results lead to an interpretation that the observed NIRS change in Δ[deoxy] level might even arise from capillary blood flow adjustment and may support the idea that changes in capillary-to-muscle O_2 delivery precisely match muscle O_2 consumption (mVO_2) [8, 9].

In contrast, ^1H magnetic resonance spectroscopy (MRS) experiments have shown that Mb does release its O_2 store, as reflected in the rising signal intensity of deoxy-Mb His F8 $N_\delta H$ during muscle contraction [10, 11]. The decreasing intracellular O_2 level implicates a widening O_2 gradient from capillary to cell to facilitate O_2 flux into the cell. In fact, Mb desaturates with rapid kinetics, consistent with a transient mismatch of O_2 supply and demand [12]. The results suggest that the observed NIRS signal contains a large Mb component.

K. Masuda, Ph.D. (✉)
Faculty of Human Sciences, Kanazawa University, Kanazawa, Ishikawa 920-1192, Japan
e-mail: masuda@ed.kanazawa-u.ac.jp

T. Jue and K. Masuda (eds.), *Application of Near Infrared Spectroscopy in Biomedicine*, Handbook of Modern Biophysics 4, DOI 10.1007/978-1-4614-6252-1_6,

Indeed, comparative NIRS and NMR studies have demonstrated that Mb contributes significantly to the observed NIRS signal (Δ[deoxy]) [13]. These results agree with a recent combined ^{1}H-NMR/NIRS study using human and canine muscles, which attributed a comparable contribution from Hb and Mb to the NIRS signal [14]. Theoretical model studies have not completely resolved this debate [8, 15].

The present chapter introduces an experimental system and observations of Mb saturation during muscle contraction to provide insight into the mechanisms regulating oxygen delivery and consumption.

6.2 Spectral Properties of Mb and Hb Ex Vivo

A solution of deoxygenated Hb and Mb has a typical red violet color, which becomes greenish upon high dilution. The absorption spectrum is invariant over the pH range in which Hb is stable and is characterized by a single broad and asymmetrical band in the visible, with a maximum at ~555 nm (for HbO_2). The Soret band in the near ultraviolet shows a maximum at about 430 nm in HbO_2 and 435 nm in MbO_2. In the ultraviolet and near infrared ($\varepsilon = 0.33$ at 760 nm for both HbO_2 and MbO_2; the absorption maximum appears in this region), the spectrum shows specific features (see Fig. 6.1). The NIR spectra of both oxy- and deoxy-Mb are quite similar to those of Hb [17] (Fig. 6.2). Although the Mb NIR spectrum shifts rightward from that of Hb, it is still very difficult to use the spectral difference to discriminate the relative contribution of Mb and Hb during in-vivo experiments.

Reynafarje proposed an algorithm to discriminate Mb and Hb signals from a mixture of Hb and Mb in blood-perfused tissue that assumes identical intensities for the HbCO α (568-nm) and β (538-nm) bands [18]. This method deconvolutes the optical spectra of Mb and Hb predicated on the assumption that the HbCO peak at 538 nm exhibits the same intensity as the peak at 568 nm (extinction coefficient for HbCO: $\varepsilon_{538\text{HbCO}} = \varepsilon_{568\text{HbCO}} = 14.7 \times 10^3\ \text{cm}^{-1}\ \text{M}^{-1}$). Therefore, with a mixture of MbCO and HbCO found in blood-perfused tissue, the optical density difference at 538 and 568 nm should yield only the Mb contribution. Dividing the intensity difference spectra by MbCO extinction coefficients at 538 and 568 nm leads to determination of MbCO concentration in the presence of Hb. The following equations describe the signal intensity difference at 538 nm (β) and 568 nm (α) to yield the concentration of MbCO (C_{MbCO}) and HbCO (C_{HbCO}):

$$\text{OD}_{538} - \text{OD}_{568} = (\varepsilon_{538\text{HbCO}} - \varepsilon_{568\text{HbCO}})C_{\text{HbCO}} + (\varepsilon_{538\text{MbCO}} - \varepsilon_{568\text{MbCO}})C_{\text{MbCO}}, \tag{6.1}$$

where OD is the optical density (absorbance or signal intensity), ε is the extinction coefficient, and C_{MbCO} and C_{HbCO} denote the concentrations of MbCO and HbCO, respectively. The Reynafarje method assumes an identical Hb extinction coefficient at 538 and 568 nm ($\varepsilon_{538\text{HbCO}} = \varepsilon_{568\text{HbCO}} = 14.7 \times 10^3$), so that the first term on the right-hand side is canceled. In addition, it assumes for Mb extinction coefficients of $\varepsilon_{538\text{MbCO}} = 14.7 \times 10^3$ and $\varepsilon_{568\text{MbCO}} = 11.8 \times 10^3\ \text{cm}^{-1}\ \text{M}^{-1}$ at 538 and 568 nm, respectively. As a consequence, the equations for C_{MbCO} and C_{HbCO} become

$$C_{\text{MbCO}} = \frac{\text{OD}_{538} - \text{OD}_{568}}{\varepsilon_{538\text{MbCO}} - \varepsilon_{568\text{MbCO}}} = \frac{\text{OD}_{538} - \text{OD}_{568}}{14.7 \times 10^3 - 11.8 \times 10^3} = (\text{OD}_{538} - \text{OD}_{568}) \times 3.45 \times 10^{-4}\text{M}, \tag{6.2}$$

$$C_{\text{HbCO}} = \frac{\text{OD}_{538} - \varepsilon_{538\text{MbCO}} \times C_{\text{MbCO}}}{\varepsilon_{538\text{HbCO}}}. \tag{6.3}$$

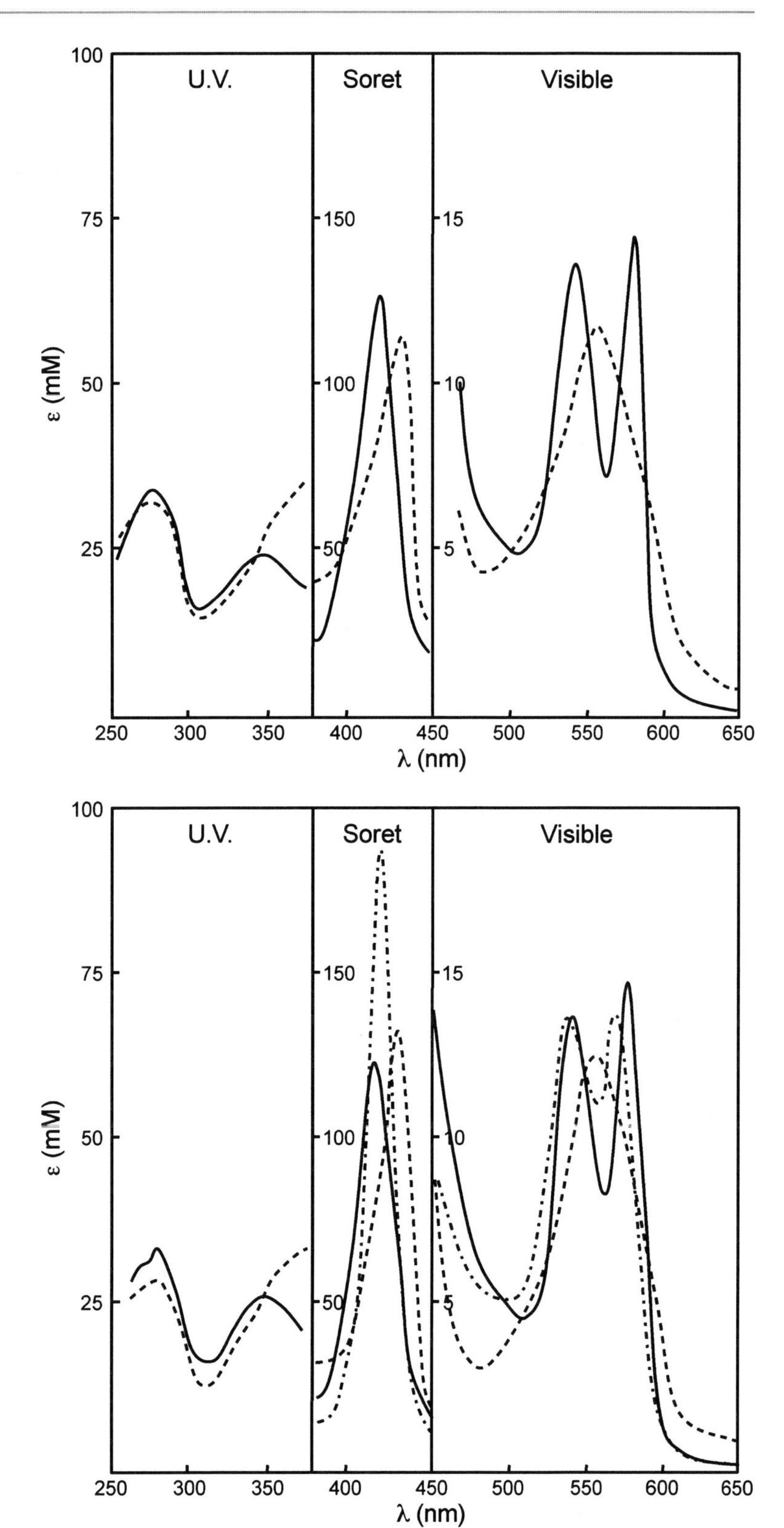

Fig. 6.1 **Spectral properties of sperm whale Mb and Hb** [16]: (*Upper*) spectrum in Mb (—: oxy, – – –: deoxy). (*Lower*) spectrum in Hb (—: oxy, – – –: deoxy, — - —: carbon monoxide), pH 7.0, 20°C

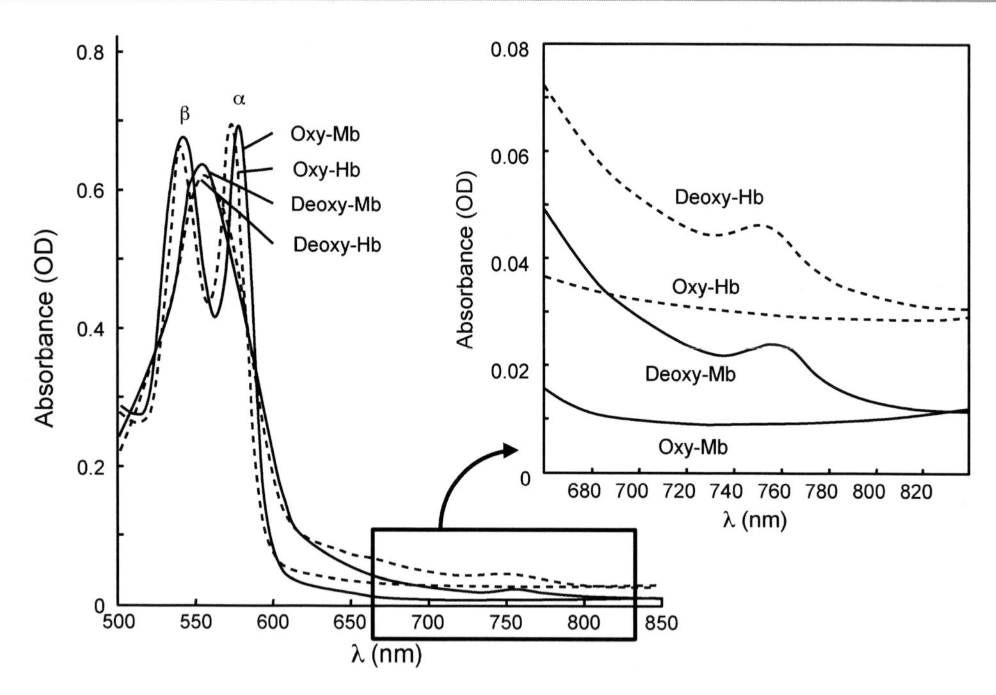

Fig. 6.2 Visible and near-infrared spectra of Hb and Mb in oxygenated and deoxygenated states [17]: (*Left*) The difference in magnitude of absorption between the visible and near-infrared spectral regions can be seen for these four chromophores. The absorbance spectra in the visible region demonstrate the subtle but measurable difference in the α and β absorption peaks between oxy-Hb and oxy-Mb. The difference between the deoxy peaks is less apparent due to their width. (*Right*) An expanded scale to show the smaller absorbance peaks in the near infrared. Although the deoxy peaks are broad, the right shift of Mb relative to hemoglobin can be seen

The potential error of the standard Reynafarje method was recently revised upward [19]. Figure 6.3 shows the optical spectra of HbCO, MbCO, and an MbCO/HbCO mixture obtained from mouse myocardium. For an HbCO solution from mouse blood, the peak maxima appear at 537 (β) and 567 nm (α) (Fig. 6.3a). For MbCO from buffer-perfused mouse myocardium, the two maxima appear at 540 (β) and 577 nm (α) (Fig. 6.3b). The spectrum of tissue homogenate obtained from unperfused myocardium containing blood shows a contribution from both Hb and Mb. The spectrum displacement is not identical to the proposed optical spectra obtained by the Reynafarje method.

The Hb absorbance maxima of different species do not always appear at the same wavelengths or with the same intensities [19]. For instance, the peaks at 538 and 568 nm can exhibit quite different intensities among species: maximal positions in mammalian Hb can vary by at least ±1 nm, and the two extinction coefficients can vary from 11.4 to 15.0 × 10^3 cm^{-1} M^{-1}. The specified absorbance values at 538 and 568 nm will thus vary as well. The Mb absorbance maxima and extinction coefficients vary widely: the MbCO β band appears from 532 to 542 nm and has extinction coefficients ranging from 11.9 to 14.8 × 10^3 cm^{-1} M^{-1} [19]. The corresponding α band can appear from 562 to 580 nm and can have an extinction coefficient between 10.6 and 12.3 × 10^3 cm^{-1} M^{-1}. Mouse Mb has β and α bands at 540 and 577 nm, respectively (Fig. 6.3b). The Reynafarje method sets MbCO extinction coefficients at 14.7 and 11.8 × 10^3 cm^{-1} M^{-1} and HbCO extinction coefficients at 14.7 and 14.7 × 10^3 cm^{-1} M^{-1} at 538 and 568 nm for all species with an assumed error of 3–10% [19].

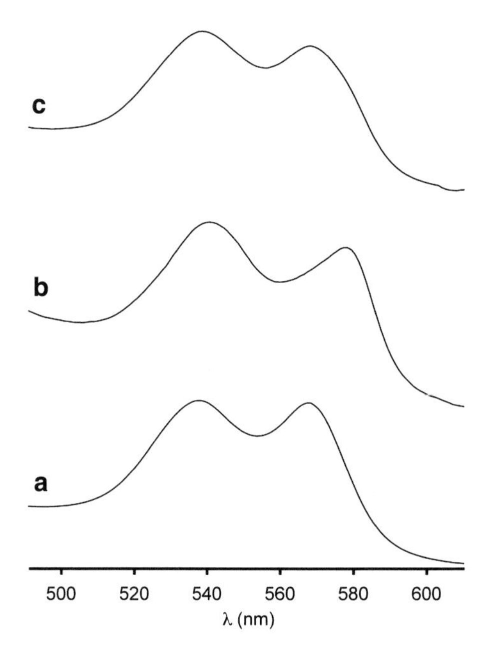

Fig. 6.3 Visible spectra of HbCO, MbCO, and MbCO/HbCO mixture [19]: (**a**) HbCO extracted from mouse blood. (**b**) MbCO extracted from buffer-perfused mouse heart. (**c**) one-to-one mixture of mouse MbCO/HbCO

6.3 Preparation of Hindlimb Perfusion

Hindlimb perfusion allows excluding the interference of Hb and is one way to extract the Mb signal from the NIR composite signal in vivo. Some investigators prepared isolated rat hindlimb as previously described [19–23]. After a midline abdominal incision, the superficial epigastric vessels were ligated. Cauterization and/or ligation occluded patent vessels, including the inferior mesenteric, pubic-epigastric, abdominal, and superior vesicle arteries, which could otherwise divert blood flow from hindlimb tissue. The abdominal wall was then incised from the pubic symphysis to the xiphoid process. The spermary, testis, and inferior mesenteric arteries and veins were ligated, and the spermaries, testes, and part of the descending colon were excised, together with contiguous adipose tissue. To perfuse only the left hindlimb, the caudal artery and internal iliac artery and vein in the right hindlimb were also ligated. A ligature was placed around the neck of the bladder, the coagulating gland, and the prostate gland. The vessels that supply the subcutaneous region were then ligated. Thereafter, the inferior epigastric, iliolumbar, and renal arteries and veins were ligated as well as the coeliac axis and portal vein [21, 23]. The inflow catheter was placed in the descending aorta, with its tip 4–5 mm proximal to the aorta bifurcation. The hypogastric trunk was initially occluded to determine whether blood flow to the trunk region could be eliminated. A ligature was then placed around the contralateral (right hindlimb) common iliac artery. Ligatures were placed around the aorta and vena cava to secure the arterial and venous catheters after perfusion had begun. Blood flow to the tail was eliminated by ligation at its base. Venous effluents were collected separately from the vena cava.

A Krebs-Henseleit buffer was employed, which consisted of (mM) 118 NaCl, 4.7 KCl, 1.2 KH_2PO_4, 1.8 $CaCl_2$, 20 $NaHCO_3$, 1.2 $MgSO_4$, and 15 parts glucose. The buffer was equilibrated with 95% O_2 + 5% CO_2 and passed through 5- and 0.45-μm Millipore filters. A circulating water bath

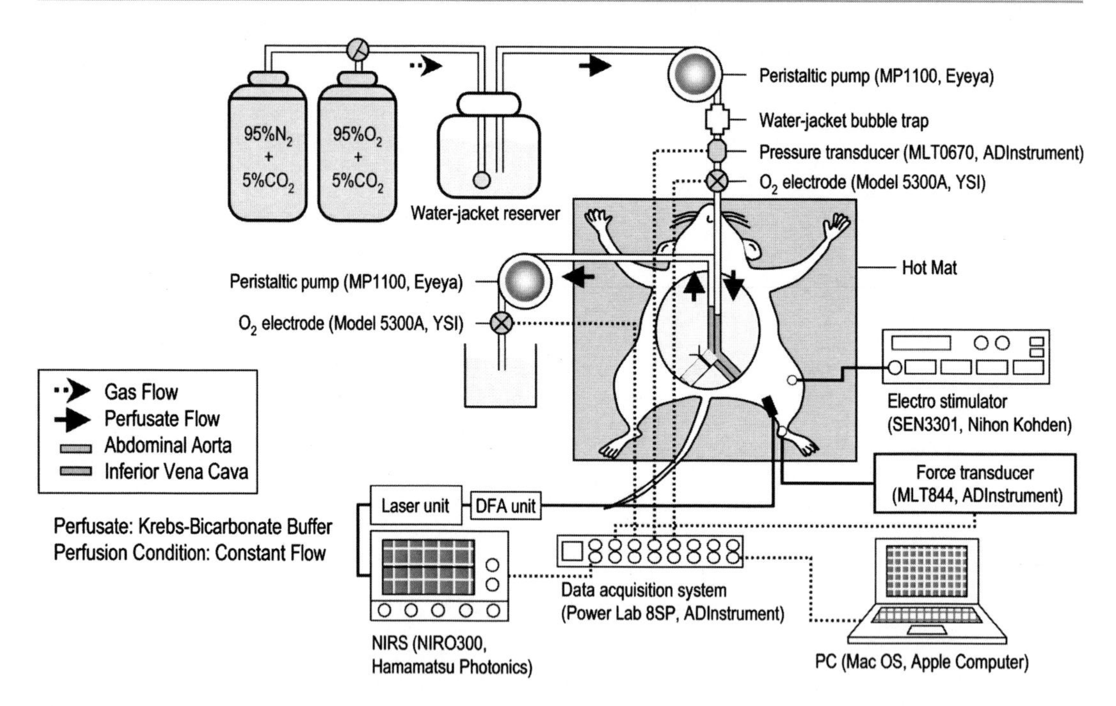

Fig. 6.4 Illustration of buffer-perfused hindquarter system: [21, 23]: Krebs bicarbonate buffer solution was bubbled with a mixing gas (95% O_2 + 5% CO_2, 37°C). After 30 min of pre-perfusion (equilibrium state), the twitch contraction started. The perfusate was infused from the abdominal artery and corrected from the abdominal vena cava

and jacketed reservoir and tubing maintained temperature at 37°C. A peristaltic pump maintained a constant, non-recirculating perfusate flow (Fig. 6.4). After cannulation of the abdominal aorta, the Krebs-Henseleit buffer containing heparin (2,000 U/l) was perfused into the hindlimb for 30 min to prevent blood clotting and to completely wash out blood from the hindlimb. During each perfusion period (~60 min), the perfusion pressure and flow rate remained constant at approximately 98–100 mmHg, which yielded a flow rate of 20–22 ml/min. Arterial PO_2 was typically 450–550 mmHg [20–31]. Under this circumstance, the hindquarter muscles would receive 0.367 ml O_2 per min from the Krebs-Henseleit buffer.

The effluent was collected from the inferior vena cava using a peristaltic pump to measure the rate of O_2 uptake. O_2 content in the influent and effluent buffers was monitored using O_2 electrodes (at 37°C). O_2 consumption (μmol/g/min) was calculated from the following equation:

$$m\dot{V}O_2 = f \times [(PO_2a - PO_2v) \times K_{O_2}] \div mW, \tag{6.4}$$

where f is the flow rate, PO_2a is the arterial partial pressure of O_2, PO_2v is the venous partial pressure of O_2, KO_2 is O_2 solubility in the buffer, and mW is muscle wet weight. The vapor pressure was 47.03 mmHg at 37°C and O_2 solubility in the buffer was 0.00135 μmol/ml/mmHg at 37°C [32].

6.4 Mb Desaturation During Skeletal Muscle Contraction

An NIRS apparatus (NIRO-300 + Detection Fiber Adapter Kit, Hamamatsu Photonics, Shizuoka, Japan) that uses single-distance continuous-wave light ($NIRS_{CWS}$) was employed to measure oxygenation of Mb [21, 23]. Based on the Beer-Lambert law, the equation for the multiple-wavelength

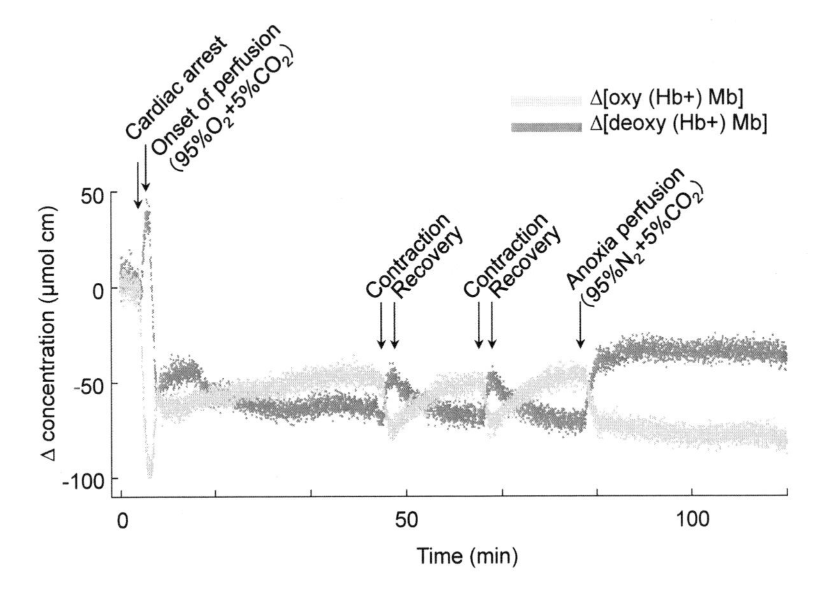

Fig. 6.5 Trends of NIR [oxy] and [deoxy] signals on hindquarter muscles during perfusion experiment: After cardiac arrest, Krebs bicarbonate buffer was perfused to the hindlimb. Then both oxy and deoxy signals showed steep decline due to removal of oxy-Hb. During equilibrium state (where only Mb was detectable), the contraction induced an increase in the deoxy signal and a decrease in the oxy signal, symmetrically. The anoxia buffer was perfused to obtain full desaturation of Mb

method can be applied to calculate changes in Δ[oxy-(Hb)/Mb], Δ[deoxy-(Hb)/Mb], or Δ[total-(Hb)/Mb]. The distance between the photodiode and LED was fixed at 10 mm. A clamp secured the toe of the foot, with the rat laid on its back. The NIRS probes were then firmly attached to the skin of the gastrocnemius muscle and fixed by clamps on both sides of the muscle. During the initial period, at least 30 s before onset of contraction, the average fluctuation of NIRS signals was adjusted to a reference value of zero. After the exercise protocol, the anoxic buffer, equilibrated with 95% N_2 + 5% CO_2 gas, was perfused for 30 min to obtain maximal Mb desaturation (Fig. 6.5). The equilibrium period with the anoxic buffer was initiated after a 5-min recovery period and after NIRS signals returned to baseline. The NIRS signal reached steady state after 30 min. The muscle then received electrical stimulation for 2 min. No further increase in Δ[deoxy-Mb] signal was evident. The final Δ[deoxy-Mb] signal intensity served as the normalization constant for 100% Mb deoxygenation.

6.5 Calculation of Intracellular PO_2 Equivalent with S_{MbO2}

Dissociation of O_2 from Mb is expressed as follows:

$$[\mathrm{Mb}] + [\mathrm{O_2}] \Leftrightarrow [\mathrm{MbO_2}],$$

$$\mathrm{K} = \frac{[\mathrm{MbO_2}]}{[\mathrm{Mb}][\mathrm{O_2}]}, \tag{6.5}$$

where $[MbO_2]$ is the concentration of oxy-Mb, [Mb] is the concentration of deoxy-Mb, $[O_2]$ is free O_2, and K is a dissociation constant. Since Mb has a single ligand with O_2, the O_2 saturation of Mb is calculated from the following:

$$S_{Mb} = \frac{[MbO_2]}{[Mb] + [MbO_2]} = \frac{K[Mb][O_2]}{[Mb] + K[Mb][O_2]}, \tag{6.6}$$

where S_{MbO_2} is the fraction of Mb saturated with O_2. Then Eq. 6.3 is converted as follows:

$$S_{MbO_2} = \frac{K[O_2]}{1 + K[O_2]} = \frac{[O_2]}{1/K + [O_2]} = \frac{PO_2}{P_{50} + PO_2}, \tag{6.7}$$

where P_{50} is the partial pressure of O_2 required to half-saturate Mb, and PO_2 is the intracellular partial pressure of O_2 equivalent to Mb saturation. Since Δ[deoxy-Mb] is measureable using the NIRS device, O_2 saturation of intracellular Mb (S_{MbO_2}) can be plotted as follows:

$$S_{MbO_2} = 100 - \%\Delta[\text{deoxy-Mb}]. \tag{6.8}$$

The Δ[deoxy-Mb] kinetic was calibrated against two different NIRS signal values: one at rest and the other during steady-state and anoxic buffer perfusion. These values correspond to an initial Mb desaturation of 10% and a final Mb desaturation of 100%, respectively (Fig. 6.6A, B). While S_{MbO2} at rest could not be determined by NIRS, the value was assumed to be 90% based on previous studies that reported S_{MbO2} at rest greater than 90% [12, 33]. Equation 6.4 can then be converted to the following equation to obtain the partial pressure of O_2 in myocytes:

$$PO_2 = \frac{S_{MbO_2}}{(1 - S_{MbO_2})} \times P_{50}. \tag{6.9}$$

The known P_{50} for Mb (e.g., 2.4 mmHg at 37°C muscle temperature [34]) would enable one to assume intracellular PO_2 to be equivalent with S_{MbO2} (Fig. 6.6B, C). The intracellular O_2 concentration ($[O_2]_{cyto}$) is then calculated as

$$[O_2]_{cyto} = PO_2 \times K_{O_2}, \tag{6.10}$$

where PO_2 is the intracellular O_2 tension in mmHg, and K_{o2} is O_2 solubility in the buffer (0.00135 μmol/ml/mmHg at 37°C [32]).

6.6 Evaluation of Mb-Associated O_2 Dynamics

Employing Eq. 6.5, the modeled S_{MbO_2} kinetics were obtained based on [deoxy-Mb] kinetics (Fig. 6.6A, B). S_{MbO_2} kinetics show no evidence for any slow component of Mb desaturation [21]. Therefore, the S_{MbO_2} plots can be fitted by the following single exponential equation to calculate the kinetics parameters using an iterative least-squares technique (Fig. 6.7):

$$S_{MbO_2} = BL + AP \times \left[1 - \exp^{-(t-TD)/\tau}\right], \tag{6.11}$$

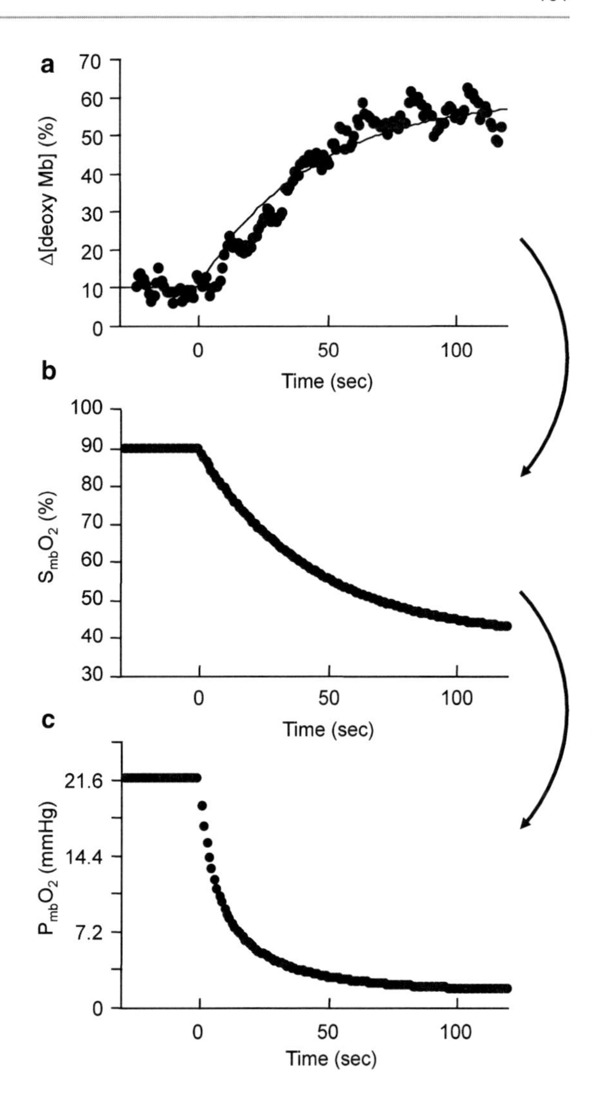

Fig. 6.6 **Conversion of NIR deoxy-Mb signal to intracellular PO_2 equivalent with S_{MbO2}** [21–23]: NIRS measures Mb deoxygenation during muscle contraction and the deoxy-Mb signal is standardized to its maximal intensity in order to express the relative trend (*graph A*). The relative change in the NIR deoxy-Mb signal can be converted and constructed to S_{MbO2} as a symmetrical curve (*graph B*). Based on the kinetics parameter of the model, S_{MbO2} kinetics was converted using Eq. 6.6

where BL is the baseline value of S_{MbO_2} kinetics, AP is the amplitude between BL and the steady-state S_{MbO_2} value during the exponential component, TD is the time delay between the start of contraction and appearance of S_{MbO_2} kinetics signals, and τ is the time constant of the kinetics of the S_{MbO_2} signal. S_{MbO_2} kinetics were modeled as an exponential function. Since $m\dot{V}O_2$ increases rapidly without time delay, the mean response time (MRT) calculated by TD + τ was used as an effective parameter of the response time for Mb desaturation at the onset of muscle contraction [35, 36].

Moreover, MRT indicates the time required to reach 63% of AP. Dividing 63% of AP by MRT yields a value for the time-dependent change in Mb desaturation. The parameter for MbO_2 release rate (f_{MbO_2}) is then calculated using the following:

$$f_{MbO_2} = \frac{0.63 \times AP}{\sqrt{MRT}} \times [Mb], \tag{6.12}$$

where (0.62 × AP/MRT) for S_{MbO2} is Mb desaturation rate (%/s), which indicates the amount of oxygen released by Mb per unit time at onset of exercise, and [Mb] is the Mb concentration in hindquarter muscles (0.12 μmol/g tissue). The MbO_2 release rate (f_{MbO_2}) is given in μmol/g/s.

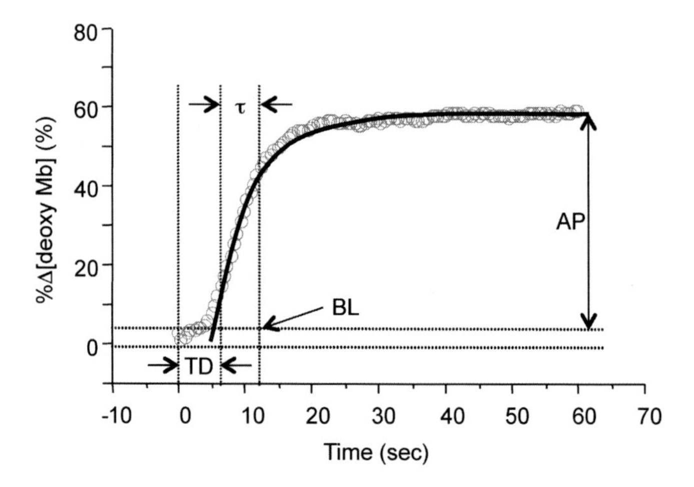

Fig. 6.7 Typical example of %Δ[deoxy-Mb] kinetics during muscle contraction: The %Δ[deoxy-Mb] kinetics at a given tension level during muscle contraction is expressed as ratios relative to the value obtained in an anoxic buffer (for animal study) or under ischemia (for human study) in order to fit the data to a model, which yields the following parameters: *AP* amplitude, *BL* baseline, *RT* response time, τ time constant, *TD* time delay

6.7 Intracellular O_2 Flux Released from Mb

The mean rate of increase in the Δ[deoxy] signal $\left(\frac{0.63 \times AP}{MRT}\right)$ reflects the Mb desaturation kinetics as well as the MbO_2 release rate to mitochondria ($[O_2]$/s). As energy demand increases with higher muscle tension, $\left(\frac{0.63 \times AP}{MRT}\right)$ increases to reflect a widening O_2 gradient [12, 36]. At the same time, it implies that the O_2 flux to mitochondria has also increased to meet energy demand. The calculated f_{MbO_2} is linearly increased with increasing $m\dot{V}O_2$, suggesting that Mb supplies the immediate source of O_2 at onset of contraction (Fig. 6.8). In either blood- or buffer-perfused muscle, the rate of change in $\left(\frac{0.63 \times AP}{MRT}\right)$ of S_{MbO_2} kinetics as a function of tension appears similar. Nevertheless, the estimated O_2 flux appears lower than previously reported values and may arise from an underestimation of MRT derived from Δ[deoxy-Mb] kinetics [12].

6.8 O_2 Gradient and O_2 Flux with Muscle Contraction

Measurement of intracellular PO_2 also provides insight into the question of whether the O_2 gradient continues to increase as energy demand rises. Studies have indicated that the NIRS Δ[deoxy] signal reflects muscle oxygenation better than the corresponding Δ[oxy] signal [37]. Under both blood and buffer perfusion, the signal intensity of the Δ[deoxy] signal of contracting muscle increases progressively with work, up to its maximal contraction rate. The O_2 gradient then continues to expand to enhance the O_2 flux as work increases, as reflected by a decrease in intracellular PO_2 (Fig. 6.9). Since blood-perfused muscle reaches a higher tension and shows an almost 50% increase in the AP of the Δ[deoxy] signal, Mb saturation should reach an even lower level [21]. In either case, Mb desaturation does not reach a plateau as tension increases to maximum contraction. The experimental evidence supports the notion that Mb continues to deoxygenate with increasing muscle activity and expand the O_2 gradient from capillary to cell space [10, 38].

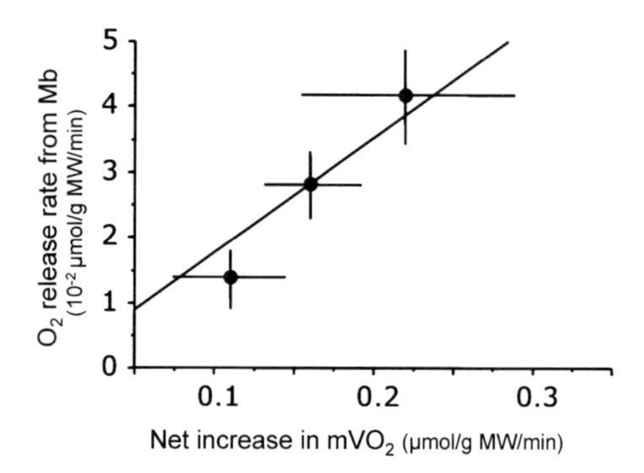

Fig. 6.8 Relationship between MbO_2 release rate and muscle oxygen consumption (net increase in $m\dot{V}O_2$) at onset of contraction [23]: Mb released O_2 immediately after onset of contraction, and the resulting oxygen release from Mb within myocyte increased with $\Delta m\dot{V}O_2$ ($r = 0.71$, $p < 0.01$). The regression line is based on mean values. Each datapoint represents mean ± SD

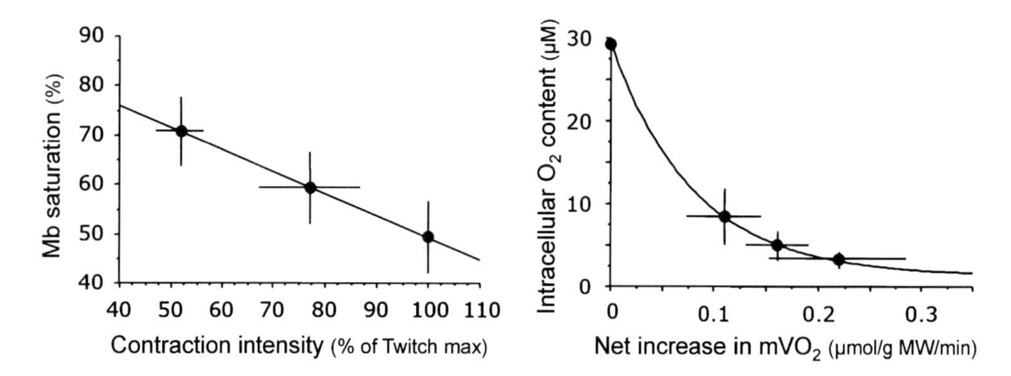

Fig. 6.9 Relationship between Mb saturation and contraction intensity during twitch contraction [23]: Mb saturation during twitch contraction decreased linearly as a function of work intensity ($r = -0.81$, $p < 0.01$). Intracellular O_2 content (μM) decreased from 29.2 at rest to 8.5 ± 3.4, 5.0 ± 1.8, and 3.3 ± 1.0 at each tension level, and continued to gradually decrease as oxygen was consumed (net increase in $m\dot{V}O_2$). Intracellular $[O_2] = 29.2 - 27.8 \times (1 - \exp(-0.08\ \Delta m\dot{V}O_2)$, $r = 0.99$). Each datapoint represents mean ± SD

The O_2 gradient will limit $m\dot{V}O_2$ in working muscle. Hypoxic buffer perfusion reduces the calculated capillary-to-cell O_2 gradient in perfuse-contracting skeletal muscle and will limit $m\dot{V}O_2$, especially when the change in O_2 gradient from rest to contraction dips below 10 μM (Ojino and Masuda, in preparation). Meanwhile, Mb O_2 saturation falls below its P_{50} (~2.0 mmHg). This evidence supports a critical PO_2 of ~2.0 mmHg for oxidative phospholyration in myocardium [39, 40].

6.9 Myoglobin Function in Muscle Cells

Standard biochemistry and physiology textbooks assert that Mb stores O_2 or facilitates O_2 transport. Although in-vitro experiments and computational models have supported the idea, in-vivo experiments have yet to present a convincing data set [41]. Marine mammal tissue contains high concentrations of Mb, which can supply O_2 during a dive [42, 43]. A correlation between Mb

Table 6.1 Estimated Mb and Hb contributions to NIRS Δ[deoxy] amplitude during contraction at a given tension level [21]

	Hb + Mb	Mb	Hb
Amplitude	26 ± 11	13 ± 6*	12 ± 5*

Unit: μM cm; means ± SD, $n = 3$, *, $p < 0.05$ vs. Hb + Mb. Hb + Mb denotes the amplitude under blood-perfused condition. Mb corresponds to the amplitude under buffer perfusion. Hb was obtained by subtracting the amplitude of Mb from Hb + Mb

concentration (O_2 supply) and species-specific aerobic activity has been shown to exist [44]. The Mb store of O_2 in rat heart, however, can prolong normal heart function only a few seconds [45]. Mice without any Mb show no impaired myocardial or skeletal muscle function [46, 47].

The advantage of Mb-facilitated O_2 diffusion within muscle cells has been predicated upon the difference between the high O_2-carrying capacity of Mb and the low solubility of O_2 [41, 48]. In-vitro Mb can actually deliver O_2 much more efficiently than faster diffusion-free O_2 [49]. Nonetheless, the in-vitro contribution of Mb remains uncertain without the presence of a value for D_{Mb}. Researchers have estimated D_{Mb} by following the diffusion of injected Mb in tissue homogenate or in myoglobin-less frog muscle [50, 51]. Others have utilized fluorescence recovery after employing a photobleaching (FRAP) technique to track injected Mb [52]. These diffusion measurements of microinjected metMb with an attached fluorophore in isolated muscle fiber have yielded a D_{Mb} of approximately 1.2×10^{-7} cm^2/s at 22°C, well below the value required to support the presence of a significant role for Mb in intracellular facilitation [53]. However, the FRAP experiments did not actually measure endogenous Mb diffusion, including an uncertain overlay of Mb oxidation/reduction kinetics, and utilized isolated muscle fibers [54].

6.10 Mb Contribution to O_2 Transport In Vivo

Comparative NIRS and NMR studies have demonstrated that Mb contributes significantly to the observed NIRS signal (Δ[deoxy]) [13]. These results agree with a recent combined ^{1}H-NMR/NIRS study that looked at human and canine muscle, which attributed 50% of the NIRS signal to Mb [14]. A comparative measurement of Δ[deoxy] found a progressively increasing NIR Δ[deoxy] signal at onset of contraction, which reached a plateau under both blood- and buffer-perfused conditions [21]. Tension-matched comparison of Δ[deoxy] level under buffer- and blood-perfused conditions indicated that Mb desaturates faster than Hb, suggesting there is an initial mobilization of the intracellular O_2 store to meet the sudden increase in energy demand at the start of contraction. Mb can also contribute approximately 50% to the NIRS signal during steady-state activity (see Table 6.1). Mathematical models with the oxy-Hb, deoxy-Hb, oxy-Mb, and deoxy-Mb components have estimated a 47% contribution of Mb to the (Δ[deoxy]) NIRS signal [15]. Further findings that Hb and Mb contribute to the NIRS signal are comparable [55]. However, other model-dependent analysis still considers Hb the predominant source of the NIRS signal [8].

[Mb] influences intracellular oxygen delivery by establishing a balanced PO_2, where the contribution from the free O_2 flux is equal to the MbO_2 flux, expressed as follows:

$$\frac{f_{MbO_2}}{f_{freeO_2}} = \frac{D_{Mb} \times [Mb]}{K_0(PO_2 + P_{50})}, \tag{6.13}$$

where f_{mbO_2} is the O_2 flux from Mb, f_{freeO_2} is the O_2 flux from dissolved O_2, D_{Mb} is the Mb diffusion coefficient, [Mb] is Mb concentration, K_0 is Krogh's diffusion constant for dissolved O_2, PO_2 is the partial pressure of O_2 at the cell surface, and P_{50} is PO_2 at half-saturated Mb [56, 57]. Experiments have depicted the relative contribution from Mb and dissolved O_2 in transporting O_2 in the cell. The straight line with respect to PO_2 depicts the contribution from dissolved O_2 diffusion and has a slope derived from a literature-reported value for Krogh's diffusion constant (K_0) of 2.52×10^{-5} ml O_2 cm^{-1} min^{-1} atm^{-1} [58]. The contribution from free O_2 flux increases linearly with PO_2. In contrast, the O_2 flux from Mb rises nonlinearly with PO_2 and depends upon [Mb], P_{50}, and D_{Mb}. Using a D_{Mb} value of 7.85×10^{-7} cm^2/s at 35°C, a P_{50} of 2 mmHg and an [Mb] of 0.2 mM, the equilibrium PO_2 diffusion is 1.7 mmHg [56, 57]. With the recent determined value of 0.26 mM in myocardium [Mb] [19], the equilibrium PO_2 rises only to 1.8 mmHg. With a resting intracellular PO_2 well above 10 mmHg and a fully saturated Mb signal at even twice the baseline workload in the heart, a 37% increase in [Mb] from 0.19 mM to 0.26 mM insignificantly alters the equilibrium PO_2 [59, 60]. Thus, only when cellular PO_2 falls below 1.7 mmHg will the role of Mb in O_2 transport dominate [56, 57]. Such a viewpoint raises questions about any simplistic interpretation of O_2 delivery enhancement following [Mb] increase with exercise training [61, 62]. Indeed, the relationship between cellular O_2 dynamics and the role of Mb requires further clarification [63].

6.11 Mb-Facilitated Diffusion and Skeletal Muscle Function

The function of Mb might differ significantly in skeletal muscle, however. Some skeletal muscle fibers can have higher concentrations of Mb than that in cardiac tissue, which would alter the equilibrium PO_2. Even though NMR detects no proximal histidyl NδH signal of deoxy-Mb in resting state, consistent with a PO_2 above 10 mmHg, there is rapid Mb desaturation at onset of contraction [10, 12, 38]. Mb desaturates within ~30 s to a steady-state level [10, 12]. For Mb-facilitated diffusion to play any role in skeletal muscle O_2 transport, intracellular PO_2 must fall below an equilibrium PO_2 of 4 mmHg, slightly above its P_{50} of 3.2 mmHg at 40°C. The dynamic desaturation of Mb solely at the beginning of muscle contraction, however, argues for a transient rather than a steady-state Mb role. No apparent O_2 limitation exists in resting muscle [10, 60, 64, 65]. At onset of exercise, a transient mismatch between O_2 supply and demand most likely arises, in contrast to research asserting a perfect match of oxygen delivery and utilization at all times [66, 67]. This mismatch requires mitochondria to draw from an immediate intracellular O_2 storage, until blood flow can adjust to meet the new intracellular O_2 demand.

6.12 Summary

The present chapter has presented the principles and an experimental apparatus for detection of intracellular O_2 dynamics employing NIRS, especially focusing on Mb saturation. Although NIRS cannot discriminate an Mb signal from a combined signal (Hb + Mb) under normal in-vivo applications, hindlimb perfusion with Hb-free buffer enables one to pick up the Mb signal, which should represent the intracellular O_2 environment during muscle contraction. The perfusion experiment provides evidence that Mb desaturates progressively and does not reach a plateau with increasing $m\dot{V}O_2$, reflecting a decrease in intracellular PO_2, Mb desaturates rapidly at onset of muscle contraction and provides O_2 to mitochondria, and Mb makes a significant (or comparable) contribution to NIR signal change. Taken together, these experimental results suggest that Mb immediately releases O_2 to mitochondria to sustain oxidative phosphorylation in contracting muscle. Further study will be necessary to clarify the molecular interaction of Mb with mitochondria and how Mb diffuses and reaches different populations of mitochondria during muscle contraction.

Acknowledgments I would like to acknowledge a grant-in-aid for Scientific Research from the Japanese Ministry of Education, Science, Sports and Culture (20680032, 23300237, KM) and partial support from the Yamaha Motor Foundation for Sports (KM). I also greatly appreciate invaluable scientific discussions with Dr. Hisashi Takakura.

Problem

6.1. NIRS measurement of Mb saturation (S_{MbO2}) at different tensions during muscle contraction shows the following:

Tension (%)	S_{MbO_2} (%)
50	70
75	59
100	49

What is the corresponding change in PO_2, given an Mb P_{50} of 2.37 at 37°C? What is the change in O_2 gradient, given a resting PO_2 of 10 mmHg? Plot out the curves. Does the change in S_{MbO2}, PO_2, and O_2 gradient show a linear relationship? What is the physiological implication in interpreting the S_{MbO2} data with respect to O_2 gradient?

Further Reading

Garry DJ, Ordway GA, Lorenz JN, Radford NB, Chin ER, Grange RW, Bassel-Duby R, Williams RS (1998) Mice without myoglobin. Nature 395:905–908

Kreutzer U, Jue T (1995) Critical intracellular O_2 in myocardium as determined by ^1H nuclear magnetic resonance signal of myoglobin. Am J Physiol 268:H1675–H1681

Lai N, Zhou H, Saidel GM, Wolf M, McCully K, Gladden LB, Cabrera ME (2009) Modeling oxygenation in venous blood and skeletal muscle in response to exercise using near-infrared spectroscopy. J Appl Physiol 106:1,858–1,874

Lin PC, Kreutzer U, Jue T (2007) Myoglobin translational diffusion in rat myocardium and its implication on intracellular oxygen transport. J Physiol 578:595–603

References

1. Popel AS (1989) Theory of oxygen transport to tissue. Crit Rev Biomed Eng 17:257–321
2. Costes F, Barthelemy JC, Feasson L, Busso T, Geyssant A, Denis C (1996) Comparison of muscle near-infrared spectroscopy and femoral blood gases during steady-state exercise in humans. J Appl Physiol 80:1345–1350
3. Seiyama A, Hazeki O, Tamura M (1988) Noninvasive quantitative analysis of blood oxygenation in rat skeletal muscle. J Biochem 103:419–424
4. Theorell H (1938) Kristallinisches myoglobin. Biochem Z 268:73–82
5. Wilson JR, Mancini DM, McCully K, Ferraro N, Lanoce V, Chance B (1989) Noninvasive detection of skeletal muscle underperfusion with near-infrared spectroscopy in patients with heart failure. Circulation 80:1668–1674
6. Bank W, Chance B (1994) An oxidative defect in metabolic myopathies: diagnosis by noninvasive tissue oximetry. Ann Neurol 36:830–837
7. McCully KK, Hamaoka T (2000) Near-infrared spectroscopy: what can it tell us about oxygen saturation in skeletal muscle? Exerc Sport Sci Rev 28:123–127
8. Harper AJ, Ferreira LF, Lutjemeier BJ, Townsend DK, Barstow TJ (2006) Human femoral artery and estimated muscle capillary blood flow kinetics following the onset of exercise. Exp Physiol 91:661–671
9. Kindig CA, Richardson TE, Poole DC (2002) Skeletal muscle capillary hemodynamics from rest to contractions: implications for oxygen transfer. J Appl Physiol 92:2513–2520
10. Mole PA, Chung Y, Tran TK, Sailasuta N, Hurd R, Jue T (1999) Myoglobin desaturation with exercise intensity in human gastrocnemius muscle. Am J Physiol 277:R173–R180
11. Richardson RS, Newcomer SC, Noyszewski EA (2001) Skeletal muscle intracellular PO_2 assessed by myoglobin desaturation: response to graded exercise. J Appl Physiol 91:2679–2685

12. Chung YR, Mole PA, Sailasuta N, Tran TK, Hurd R, Jue T (2005) Control of respiration and bioenergetics during muscle contraction. Am J Physiol Cell Physiol 288:C730–C738
13. Tran TK, Sailasuta N, Kreutzer U, Hurd R, Chung Y, Mole P, Kuno S, Jue T (1999) Comparative analysis of NMR and NIRS measurements of intracellular PO_2 in human skeletal muscle. Am J Physiol 276:R1682–R1690
14. Nioka S, Wang DJ, Im J, Hamaoka T, Wang ZJ, Leigh JS, Chance B (2006) Simulation of Mb/Hb in NIRS and oxygen gradient in the human and canine skeletal muscles using H-NMR and NIRS. Adv Exp Med Biol 578:223–228
15. Hoofd L, Colier W, Oeseburg B (2003) A modeling investigation to the possible role of myoglobin in human muscle in near-infrared spectroscopy (NIRS) measurements. Adv Exp Med Biol 530:637–643
16. Antonini E, Brunori M (eds) (1971) Hemoglobin and myoglobin in their reactions with ligands. North-Holland, Amsterdam
17. Schenkman KA, Marble DR, Feigl EO, Burns DH (1999) Near-infrared spectroscopic measurement of myoglobin oxygen saturation in the presence of hemoglobin using partial least-squares analysis. Appl Spectrosc 53:325–331
18. Reynafarje B (1963) Simplified method for the determination of myoglobin. J Lab Clin Med 61:138–145
19. Masuda K, Truscott K, Lin PC, Kreutzer U, Chung Y, Sriram R, Jue T (2008) Determination of myoglobin concentration in blood-perfused tissue. Eur J Appl Physiol 104:41–48
20. Lewis SB, Schultz TA, Westbie DK, Gerich JE, Wallin JD (1977) Insulin-glucose dynamics during flow-through perfusion of the isolated rat hindlimb. Horm Metab Res 9:190–195
21. Masuda K, Takakura H, Furuichi Y, Iwase S, Jue T (2010) NIRS measurement of O_2 dynamics in contracting blood and buffer perfused hindlimb muscle. Adv Exp Med Biol 662:323–328
22. Seiyama A (2006) Virtual cooperativity in myoglobin oxygen saturation curve in skeletal muscle in vivo. Dyn Med 5:3
23. Takakura H, Masuda K, Hashimoto T, Iwase S, Jue T (2010) Quantification of myoglobin deoxygenation and intracellular partial pressure of O_2 during muscle contraction during haemoglobin-free medium perfusion. Exp Physiol 95:630–640
24. Constantin-Teodosiu D, Baker D, Constantin D, Greenhaff P (2008) PPARdelta agonism inhibits skeletal muscle PDC activity, mitochondrial ATP production and force generation during prolonged contraction. J Physiol 587:231–239
25. Raney MA, Turcotte LP (2006) Regulation of contraction-induced FA uptake and oxidation by AMPK and ERK1/2 is intensity dependent in rodent muscle. Am J Physiol Endocrinol Metab 291:E1220–E1227
26. Raney MA, Turcotte LP (2008) Evidence for the involvement of CaMKII and AMPK in Ca^{2+}-dependent signaling pathways regulating FA uptake and oxidation in contracting rodent muscle. J Appl Physiol 104:1366–1373
27. Ruderman NB, Houghton CR, Hems R (1971) Evaluation of the isolated perfused rat hindquarter for the study of muscle metabolism. Biochem J 124:639–651
28. Shiota M, Golden S, Katz J (1984) Lactate metabolism in the perfused rat hindlimb. Biochem J 222:281–292
29. Shiota M, Sugano T (1986) Characteristics of rat hindlimbs perfused with erythrocyte- and albumin-free medium. Am J Physiol 251:C78–C84
30. Seiyama A, Kosaka H, Maeda N, Shiga T (1996) Effect of hypothermia on skeletal muscle metabolism in perfused rat hindlimb. Cryobiology 33:338–346
31. Bonen A, Clark MG, Henriksen EJ (1994) Experimental approaches in muscle metabolism: hindlimb perfusion and isolated muscle incubations. Am J Physiol Endocrinol Metab 266:E1–E16
32. Philip LA, Dorothy SD (1971) Respiration and circulation. FASEB, Bethesda
33. Richardson RS, Duteil S, Wary C, Wray DW, Hoff J, Carlier PG (2006) Human skeletal muscle intracellular oxygenation: the impact of ambient oxygen availability. J Physiol 571:415–424
34. Schenkman KA, Marble DR, Burns DH, Feigl EO (1997) Myoglobin oxygen dissociation by multi-wavelength spectroscopy. J Appl Physiol 82:86–92
35. Behnke BJ, Kindig CA, McDonough P, Poole DC, Sexton WL (2002) Dynamics of microvascular oxygen pressure during rest-contraction transition in skeletal muscle of diabetic rats. Am J Physiol Heart Circ Physiol 283: H926–H932
36. Rossiter HB, Ward SA, Doyle VL, Howe FA, Griffiths JR, Whipp BJ (1999) Inferences from pulmonary O_2 uptake with respect to intramuscular [phosphocreatine] kinetics during moderate exercise in humans. J Physiol 518(Pt 3): 921–932
37. Deblasi RA, Ferrari M, Natali A, Conti G, Mega A, Gasparetto A (1994) Noninvasive measurement of forearm blood-blow and oxygen consumption by near-infrared spectroscopy. J Appl Physiol 76:1388–1393
38. Richardson RS, Noyszewski EA, Kendrick KF, Leigh JS, Wagner PD (1995) Myoglobin O_2 desaturation during exercise: evidence of limited O_2 transport. J Clin Invest 96:1916–1926
39. Kreutzer U, Jue T (1995) Critical intracellular O_2 in myocardium as determined by ^{1}H nuclear magnetic resonance signal of myoglobin. Am J Physiol 268:H1675–H1681

40. Kreutzer U, Wang DS, Jue T (1992) Observing the ^{1}H NMR signal of the myoglobin Val-E11 in myocardium: an index of cellular oxygenation. Proc Natl Acad Sci USA 89:4731–4733
41. Wittenberg BA, Wittenberg JB (1989) Transport of oxygen in muscle. Annu Rev Physiol 51:857–878
42. Guyton GP, Stanek KS, Schneider RC, Hochachka PW, Hurford WE, Zapol DG, Liggins GC, Zapol WM (1995) Myoglobin saturation in free-diving Weddell seals. J Appl Physiol 79:1148–1155
43. Ponganis PJ, Kreutzer U, Sailasuta N, Knower T, Hurd R, Jue T (2002) Detection of myoglobin desaturation in *Mirounga angustirostris* during apnea. Am J Physiol Regul Integr Comp Physiol 282:R267–R272
44. Wittenberg JB, Wittenberg BA (2003) Myoglobin function reassessed. J Exp Biol 206:2011–2020
45. Chung Y, Jue T (1996) Cellular response to reperfused oxygen in the postischemic myocardium. Am J Physiol 271: H687–H695
46. Garry DJ, Ordway GA, Lorenz JN, Radford NB, Chin ER, Grange RW, Bassel-Duby R, Williams RS (1998) Mice without myoglobin. Nature 395:905–908
47. Godecke A, Flogel U, Zanger K, Ding Z, Hirchenhain J, Decking UK, Schrader J (1999) Disruption of myoglobin in mice induces multiple compensatory mechanisms. Proc Natl Acad Sci USA 96:10495–10500
48. Wittenberg JB (1970) Myoglobin-facilitated oxygen diffusion: role of myoglobin in oxygen entry into muscle. Physiol Rev 50:559–636
49. Johnson RL, Heigenhauser GJF, Hsia CCW, Jones NL, Wagner PD (1996) Determinants of gas exchange and acid–base balance during exercise. In: Rowell LB, Shepher JT (eds) Regulation and integration of multiple systems. Oxford University Press, New York, pp 515–584
50. Baylor SM, Pape PC (1988) Measurement of myoglobin diffusivity in the myoplasm of frog skeletal muscle fibres. J Physiol 406:247–275
51. Papadopoulos S, Endeward V, Revesz-Walker B, Jurgens KD, Gros G (2001) Radial and longitudinal diffusion of myoglobin in single living heart and skeletal muscle cells. Proc Natl Acad Sci USA 98:5904–5909
52. Verkman AS (2003) Diffusion in cells measured by fluorescence recovery after photobleaching. Methods Enzymol 360:635–648
53. Riveros-Moreno V, Wittenberg JB (1972) The self-diffusion coefficients of myoglobin and hemoglobin in concentrated solutions. J Biol Chem 247:895–901
54. Groebe K (1995) An easy-to-use model for O_2 supply to red muscle: validity of assumptions, sensitivity to errors in data. Biophys J 68:1246–1269
55. Lai N, Zhou H, Saidel GM, Wolf M, McCully K, Gladden LB, Cabrera ME (2009) Modeling oxygenation in venous blood and skeletal muscle in response to exercise using near-infrared spectroscopy. J Appl Physiol 106:1858–1874
56. Lin PC, Kreutzer U, Jue T (2007) Anisotropy and temperature dependence of myoglobin translational diffusion in myocardium: implication for oxygen transport and cellular architecture. Biophys J 92:2608–2620
57. Lin PC, Kreutzer U, Jue T (2007) Myoglobin translational diffusion in rat myocardium and its implication on intracellular oxygen transport. J Physiol 578:595–603
58. Bentley TB, Meng H, Pittman RN (1993) Temperature dependence of oxygen diffusion and consumption in mammalian striated muscle. Am J Physiol 264:H1825–H1830
59. Kreutzer U, Mekhamer Y, Chung Y, Jue T (2001) Oxygen supply and oxidative phosphorylation limitation in rat myocardium in situ. Am J Physiol Heart Circ Physiol 280:H2030–H2037
60. Zhang J, Murakami Y, Zhang Y, Cho YK, Ye Y, Gong G, Bache RJ, Uğurbil K, From AH (1999) Oxygen delivery does not limit cardiac performance during high work states. Am J Physiol 277:H50–H57
61. Hickson RC (1981) Skeletal muscle cytochrome c and myoglobin, endurance, and frequency of training. J Appl Physiol 51:746–749
62. Masuda K, Kano Y, Nakano H, Inaki M, Katsuta S (1998) Adaptations of myoglobin in rat skeletal muscles to endurance running training: effects of intensity, duration, and period of training. Jpn J Phys Fitness Sports Med 47:561–571
63. Kindig CA, Kelley KM, Howlett RA, Stary CM, Hogan MC (2003) Assessment of O_2 uptake dynamics in isolated single skeletal myocytes. J Appl Physiol 94:353–357
64. Kreutzer U, Mekhamer Y, Tran TK, Jue T (1998) Role of oxygen in limiting respiration in the in situ myocardium. J Mol Cell Cardiol 30:2651–2655
65. Mancini DM, Wilson JR, Bolinger L, Li H, Kendrick K, Chance B, Leigh JS (1994) In vivo magnetic resonance spectroscopy measurement of deoxymyoglobin during exercise in patients with heart failure: demonstration of abnormal muscle metabolism despite adequate oxygenation. Circulation 90:500–508
66. Behnke BJ, Kindig CA, Musch TI, Koga S, Poole DC (2001) Dynamics of microvascular oxygen pressure across the rest-exercise transition in rat skeletal muscle. Respir Physiol 126:53–63
67. Kindig CA, Howlett RA, Hogan MC (2003) Effect of extracellular PO_2 on the fall in intracellular PO_2 in contracting single myocytes. J Appl Physiol 94:1964–1970

Muscle Oxygen Saturation Measurements in Diving Mammals and Birds Using NIRS

7

Cassondra L. Williams and Paul J. Ponganis

7.1 Introduction

Near-infrared spectroscopy (NIRS) principles have been applied to study muscle oxygen (O_2) saturation in diving mammals and birds. Custom-made NIRS instruments and probes were developed to measure myoglobin (Mb) O_2 saturation levels in two studies conducted in Antarctica at isolated dive holes on the sea ice of McMurdo Sound on freely diving Weddell seals (*Leptonychotes weddellii*) [1] and on freely diving emperor penguins (*Aptenodytes forsteri*) [2]. A third study used a modified, commercially available NIRS instrument (Vander Niroscope) to measure muscle O_2 levels in trained harbor seals [3]. In that study researchers trained harbor seals for simulated dives using a helmet that filled with water for specific time periods.

The primary focus of the present chapter will be on the application of NIRS principles to the development and use of custom-made instruments and the methodology of those first two studies. The third study will be discussed as to how the instrument was applied to use in trained seals.

7.2 Background

Air-breathing marine vertebrates, such as seals and penguins, spend much of their time foraging or migrating at sea and can remain underwater for several minutes to almost 2 h [4, 5]. One of the hallmarks of diving birds and mammals is an increased Mb concentration in locomotory muscles. Mb concentrations of some marine animals are more than tenfold higher than that observed in terrestrial animals [6]. For example, Mb concentration in the hooded seal is 8.6 g 100 g^{-1}, while the Mb concentration in sheep is 0.6 g 100 g^{-1}. The increased Mb concentration provides a significant O_2 store in the muscle. When forcibly submerged, air-breathing marine vertebrates exhibit a series of physiological responses, collectively called the dive response [7, 8]. During forced submersion

C.L. Williams, Ph.D. (✉)
Department of Ecology and Evolutionary Biology, University of California Irvine, 321 Steinhaus, 92697-2525 Irvine, CA, USA
e-mail: cassondw@uci.edu

P.J. Ponganis, Ph.D., M.D.
Center for Marine Biomedicine and Biotechnology, Scripps Institution of Oceanography, University of California San Diego, La Jolla, CA 92093-0204, USA
e-mail: pponganis@ucsd.edu

T. Jue and K. Masuda (eds.), *Application of Near Infrared Spectroscopy in Biomedicine*, Handbook of Modern Biophysics 4, DOI 10.1007/978-1-4614-6252-1_7,

studies, diving animals experience a severe bradycardia (decrease in heart rate), vasoconstriction, and circulatory isolation of peripheral organs [7, 8], including skeletal muscle. These responses reduce O_2 consumption to preserve O_2 for vital organs such as the heart and brain. However, the physiological adjustments, particularly in muscle, that occur in voluntarily diving marine birds and mammals are not well understood.

Understanding the depletion of the muscle O_2 store is important in determining the physiological basis of the aerobic dive limit (ADL), which is widely used to interpret diving physiology and ecology studies of air-breathing marine vertebrates. The ADL is defined as the dive duration at which post-dive blood lactate levels begin to increase above resting values [9]. The increase in blood lactate is believed to be secondary to (1) depletion of one of the three O_2 stores (respiratory, blood, or muscle), (2) subsequent anaerobic metabolism, and (3) washout of lactate from a site during the surface period [9–11]. Isolation and then depletion of the Mb-O_2 store during forced submersion studies has led to the hypothesis that the muscle O_2 store is the trigger for the ADL. However, efforts to study muscle O_2 store depletion and to determine whether locomotory muscles are isolated from the circulation in voluntarily diving animals have been hindered by the technical complexities of measuring muscle O_2 in a moving, submerged animal.

In a forcibly submerged and immobilized harbor seal, Mb-O_2 saturation was measured by serial biopsy sampling of muscle before, during, and after submersion [7]. Muscle appeared isolated from the circulation and muscle O_2 depletion was linear until near complete depletion at 10 min. However, forced submersion models elicit more severe physiological responses since the animal does not know how long it will be until it is able to breathe again. Studies of freely diving or trained animals have demonstrated less severe physiological responses, such as a more moderate bradycardia [12–21].

NMR has also been used to examine muscle O_2 depletion during spontaneous breath holds of a northern elephant seal [22, 23]. Northern elephant seals are exceptional divers that regularly dive for 10–30 min, with maximum dive durations over 100 min [24–26]. While on land these seals experience sleep apnea, with breath holds as long as 10–20 min [27]. Using NMR, researchers examined Mb-O_2 during apneic and eupneic periods. In this study, the seal was allowed to fall asleep in an NMR magnet and Mb-O_2 was measured during natural apneas that occurred over the experimental period [23]. Mb-O_2 rapidly declined during apneic periods to 80% of pre-apneic levels and then recovered within 1 min after the end of the breath hold [23].

Studies using these techniques provided information on the depletion of O_2 during extreme conditions (forced submersions) or during natural breath holds. However, because these animals were not moving, the effect of muscle workload during measurements was necessarily excluded in these studies. In addition, these techniques cannot be used on freely diving animals. The application of NIRS principles to small, submersible instruments has allowed for measurement muscle O_2 in freely diving, actively swimming marine animals.

7.3 Development of Novel Instruments Using NIRS

7.3.1 NIR Reflectance Spectroscopy and Mb Absorption Spectra

The NIRS recorders employed in these studies used reflectance spectroscopy. There is no noninvasive method to record NIR light that is transmitted through the large muscles of wild, freely diving animals. In this application of reflectance spectroscopy, NIR light is transmitted directly on the muscle tissue. The absorption of transmitted NIR light by Mb is measured from the reduction in intensity of reflected light. The attenuation in light intensity is proportional to absorption by Mb. This attenuation is relative to the NIR absorption spectra of deoxygenated Mb (deoxy-Mb) and oxygenated Mb (oxy-Mb), which were used to obtain relative levels of Mb-O_2 saturation.

The absorption spectra for oxy-Mb and deoxy-Mb have distinct differences in absorption at approximately 760 nm. Deoxy-Mb has a higher absorption extinction coefficient, absorbing a much larger portion of light, at 760 nm than oxy-Mb (see Chap. 1). The increased attenuation results in a lower reflectance signal for deoxy-Mb at 760 nm. Thus, NIRS can be used to measure muscle O_2 saturation because of the different absorption extinction coefficients of deoxy-Mb and oxy-Mb at the 760-nm wavelength. At 800 nm there is an isosbestic point – where deoxy-Mb and oxy-Mb have the same absorption. The ratio of the reflected light at 760 and 800 nm is linearly correlated with the O_2 saturation level of Mb [1, 2, 28, 29].

The oxy-Mb and deoxy-Mb absorption spectra in the NIR range are essentially the same as the oxy-Hb and deoxy-Hb absorption spectra. As a result, the basic principles of using NIRS to measure blood O_2 saturation apply to measuring muscle O_2 saturation. However, because Hb and Mb have nearly the same absorption spectra in the NIR range, distinguishing between the absorption of Hb and Mb in situ is difficult, complicating interpretation of NIRS signals in muscle O_2 studies (see Sect. 7.3.4 below.)

7.3.2 Linearity Validation and NIRS Signal Calibration

Measurements from instruments using the NIRS principles described above provide relative changes in O_2 saturation values only if the NIRS measurements and Mb-O_2 saturation values are linearly correlated. Thus, the linearity of the relationship between the NIRS instrument measurements and different Mb-O_2 saturations must be verified for NIRS instruments using reflectance spectroscopy. Because it is difficult to alter muscle O_2 levels in vitro to specific O_2 saturation percentages, experiments to validate NIRS instrumentation for measuring muscle O_2 can be performed using blood (Hb) as a proxy for muscle (Mb), since Hb and Mb have essentially identical absorption spectra in the NIR range.

If a linear relationship is confirmed, a two-point calibration curve can be made by determining the NIRS instrument readings at the zero and 100% O_2 saturation levels. NIRS instrument readings can then be converted to O_2 saturation percentages using linear interpolation.

7.3.3 Implant Site

Although one of the benefits of NIRS is that it is a noninvasive method for determining blood O_2 saturation, this does not apply to diving animals. Measurements of muscle O_2 saturation must be done invasively, as working with marine animals presents potentially confounding effects, including: (1) a thick subcutaneous adipose tissue layer, (2) feathers, (3) skin, and (4) Hb absorption. Marine mammals have large layers of subcutaneous adipose tissue (e.g., 3–5 cm adipose layer for Weddell seals, [1]) that the NIR light would have to penetrate before reaching the muscle layer. Marine birds, such as penguins, have a thick plumage that would also likely cause significant noise in NIRS signals. In addition, the transmission properties of the skin of marine birds and mammals have not been quantified. Since only minimal in-situ verification is possible when working with wild animals that are to be returned to their natural habitat after experiments, quantifying these confounding effects would be particularly difficult.

Finally, as described above, it is difficult to distinguish between signals for Mb-O_2 and Hb-O_2. However, when probes are implanted directly on the muscle surface, highly vascularized areas can be avoided [1, 2]. Thus, for all applications on diving animals to date, NIRS probes have been implanted subcutaneously [1–3].

7.3.4 Experimental Evaluation of Instrument

Even when probes are implanted in areas free of blood vessels, NIRS signals can include both Hb-O_2 and Mb-O_2 saturation data, and the relative contributions of blood flow, Mb, and Hb to the NIRS signals cannot be distinguished in the collected data [29–31]. Thus, in order to assess the potential influences of blood flow, Mb, and Hb on the NIRS signals, experiments can be done with diving animals under anesthesia. For these experiments, the probe is attached to the diving animal as in the study protocol and then experimental manipulations of Hb concentration and saturation are conducted to assess any influence on the NIRS signals. See Sects. 7.4.1 and 7.4.2 for discussion of specific instrument evaluation experiments.

7.4 Summary of NIRS Studies of Diving Mammals and Birds

7.4.1 Muscle O_2 Depletion in Weddell Seals

In the first study to measure muscle O_2 in a freely diving animal, a dual-wavelength NIR spectrophotometer was developed to measure Mb-O_2 saturation in Weddell seals [1]. The probe consisted of two laser diodes with wavelengths of 753–755 and 812–814 nm and two photodetectors mounted 1 cm below and parallel to the laser diodes. The probe was potted with optical cement with sutures attached to the probe during the curing process to allow for easy attachment [1].

To verify the linear relationship, the instrument was tested using sheep blood as a proxy for muscle tissue. The blood O_2 was adjusted to various O_2 tensions during the experiment, and the NIRS probe was attached to the wall of the vessel containing the blood. NIRS readings and blood samples were taken, and the O_2 saturation percentage of the samples was determined using blood gas analysis and the sheep O_2-Hb dissociation curve. Regression analysis confirmed a linear relationship between the NIRS readings and O_2 saturation percentages ($r^2 = 0.91$) [1].

The NIRS probe was implanted subcutaneously on the surface of the latissimus dorsi of five subadult Weddell seals in Antarctica. Although muscles of the longissimus dorsi–iliocostalis muscle complex are probably the primary locomotory muscles for swimming in seals, the latissimus dorsi was readily accessible [1, 32, 33].

In one seal, to assess the potential influence of Hb saturation on NIRS signals, arterial saturation was measured, concurrently with NIRS muscle oxygenation measures, with an oxymetric, fiber optic Swan-Ganz catheter inserted in the aorta while the seal recovered from anesthesia. The NIRS signals suggested muscle saturation followed the same pattern as arterial saturation during apneic periods [1]. However, after administration of epinephrine to induce vasoconstriction, the NIRS signals indicated muscle O_2 did not follow the same desaturation pattern as arterial O_2. It was concluded that muscle O_2 stores were significantly isolated from central circulation during vasoconstriction under these conditions and that, in this situation, NIRS signals primarily reflected Mb saturation [1].

After recovery from the implant procedure, the seals were released into an isolated dive hole and allowed to dive freely. The resulting NIRS recordings were converted to O_2 saturation percentage using a two-point calibration. A 100% Mb-O_2 saturation was assumed at the resting

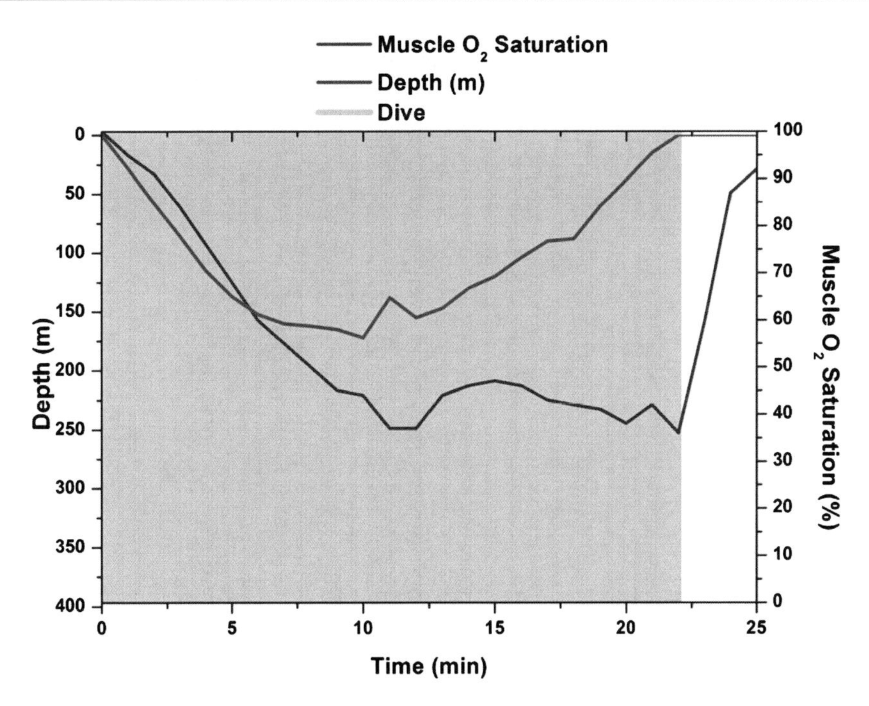

Fig. 7.1 Muscle O_2 saturation and depth profile for a 22-min, 250-m dive by a Weddell seal. Muscle O_2 fell to about 35% in the first 10 min and then leveled off until the end of the dive (Adapted from Guyton et al. [1])

baseline in the Weddell seal. A zero percent saturation value was obtained using a zero point calibration experiment. The zero calibration experiment consisted of excising the portion of the muscle that was directly beneath, and still attached to, the probe, and then recording the desaturation of the muscle until an asymptote was obtained. The NIRS reading at that point represented zero percent O_2 saturation.

The results indicated that the Mb-O_2 store was partially consumed during diving, but was never completely desaturated during any dives [1]. During long dives (>17 min), the Mb-O_2 store was desaturated by approximately 60%. In two seals, there appeared to be some resaturation of Mb during the dive. The results suggested the Mb-O_2 store was supplemented by the blood O_2 store during dives (Fig. 7.1) [1]. However, interpretation of these results was complicated by the fact that in the presence of muscle blood flow during anesthesia the NIRS signals were influenced by changes in arterial saturation [1]. The contribution of arterial saturation to the muscle NIRS signals was not quantified in this study.

7.4.2 Muscle O_2 Depletion in Emperor Penguins

In the most recent study, Williams and colleagues measured muscle O_2 depletion in emperor penguins diving at an isolated hole in Antarctica [2]. They developed an NIRS recorder small enough for use in diving emperor penguins and implanted the probe on the pectoral muscle, the primary locomotory muscle during emperor penguin dives [2].

The custom-built NIRS instrument included a microprocessor-based recorder and an implantable probe (Fig. 7.2). The probe was a small circuit board with two surface-mount LEDs with wavelengths of 760 and 810 nm positioned 3 mm apart. Two photodiodes were placed 3 mm on the outer side of

Fig. 7.2 The NIRS instrument from the Williams et al. study [2]. Probe shown attached to microprocessor. Black underwater housing case is shown next to instrument (From Williams et al. [2])

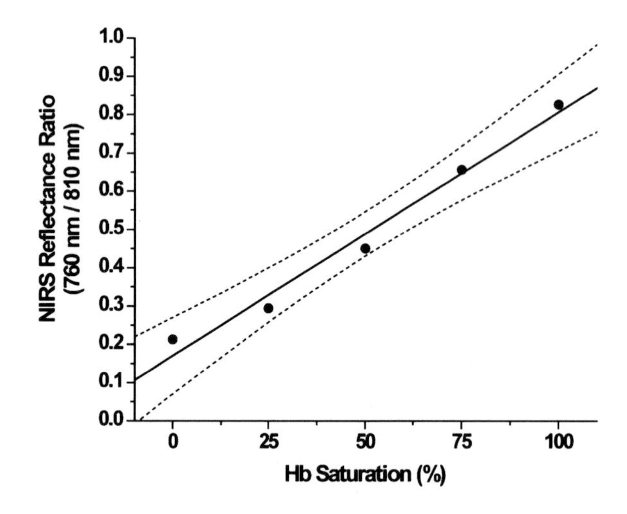

Fig. 7.3 The NIRS signal readings from one of the probes versus hemoglobin (Hb) saturation percentages show a strong linear relationship ($r^2 = 0.98$; $p = 0.0012$; *dashed lines* reflect 95% confidence intervals) (From Williams et al. study [2])

each LED. Probes were sealed with an optical epoxy. The microprocessor recorder was contained in an underwater housing unit during diving experiments (Fig. 7.2) [2].

The linear relationship between the NIRS recorder and O_2 saturation was verified using rat blood. The probe was placed in rat blood, the PO_2 of which was varied using a tonometer and a volumetric mixing technique [2]. Simultaneously, NIRS readings were taken and O_2 saturation percentages obtained from blood samples using a blood gas analyzer and the rat O_2-Hb dissociation curve [2]. The values obtained from the NIRS instrument and the O_2 saturation values from blood gas analysis were plotted to verify a linear relationship ($r^2 = 0.98$) (Fig. 7.3) [2].

In the field study in Antarctica, under anesthesia, probes were sutured subcutaneously to the surface of the pectoral muscle of penguins [2]. In two penguins, evaluation experiments were done under anesthesia to assess the influence of Hb on NIRS signals. The potential effects of muscle blood flow (i.e., the amount of Hb in muscle) on NIRS signals were evaluated by injecting alpha and beta sympathomimetic agents and monitoring any changes. Blood pressure and heart rate changed

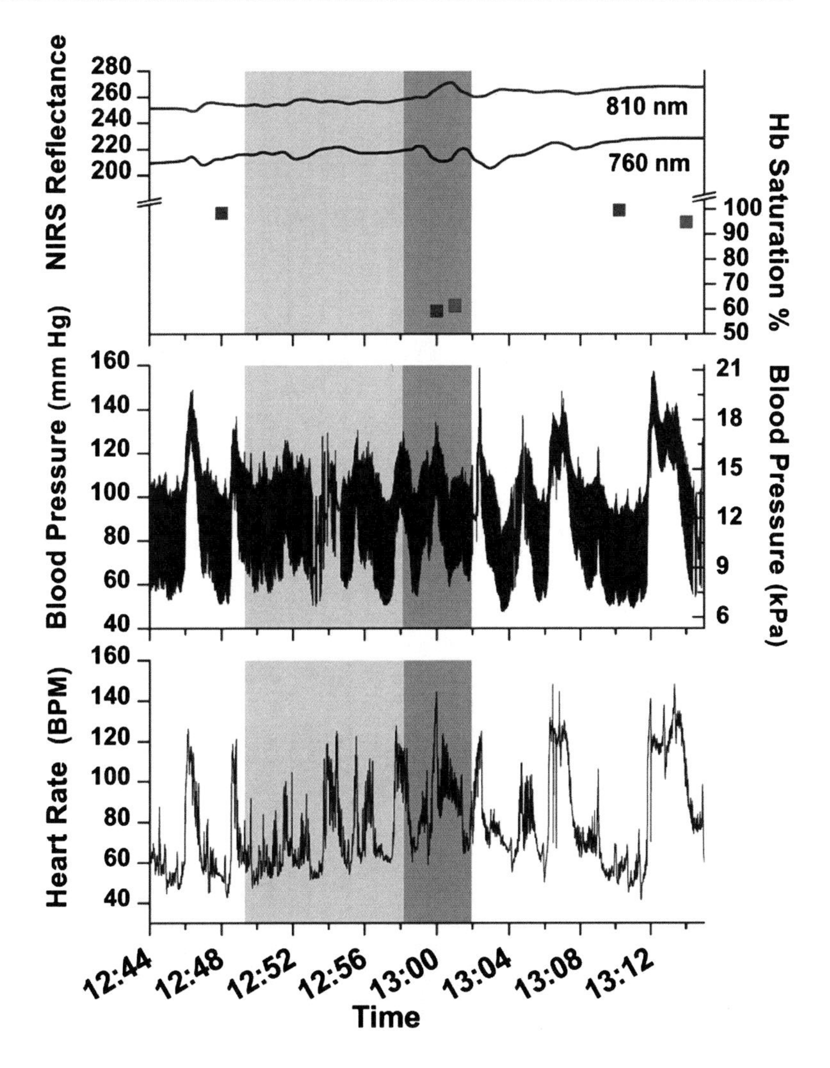

Fig. 7.4 Excerpt of instrument evaluation experiment showing NIRS reflectance, Hb saturation, blood pressure, and heart rate of an anesthetized penguin. Inspired oxygen fraction (F_{IO2}) was reduced during the experiment. *Maroon line*, 810-nm reflectance signal; *black line*, 760-nm reflectance signal; *red squares*, arterial Hb saturation; *blue squares*, venous Hb saturation; *light-gray shaded area*. $F_{IO2} = 50\%$; *dark-gray shaded area*, $F_{IO2} = 20\%$ (From Williams et al. [2])

appropriately after administration of the vasoactive drugs during the experiment, but no significant changes in NIRS signals were observed [2]. In a separate experiment, the effect of Hb saturation was evaluated by reducing the O_2 level of the administered gas. Despite an approximate 40% reduction of arterial and venous Hb saturation, only small and inconsistent changes occurred in NIRS signals (Fig. 7.4). It was concluded that the NIRS signals primarily reflect Mb saturation in penguin muscle.

After recovery from the implant procedure, penguins dived freely for 1–2 days. The NIRS data collected were converted to O_2 saturation using a two-point calibration. A 100% Mb-O_2 saturation was assumed before dives in emperor penguins. A zero percent saturation value was obtained using a zero-point calibration experiment. A portion of the muscle directly beneath the probe was excised, placed in 100% nitrogen, and the subsequent desaturation of the muscle was recorded until an asymptote was obtained, and no further changes in the signals occurred. The NIRS reading at that point was considered to represent zero percent O_2 saturation [2].

The results revealed that complete Mb-O_2 desaturation occurred in some dives and suggested muscle was essentially isolated from the circulation during these dives (Fig. 7.5a). This depletion pattern (Fig. 7.5a) suggested that the muscle was isolated from the circulation during the entire dive,

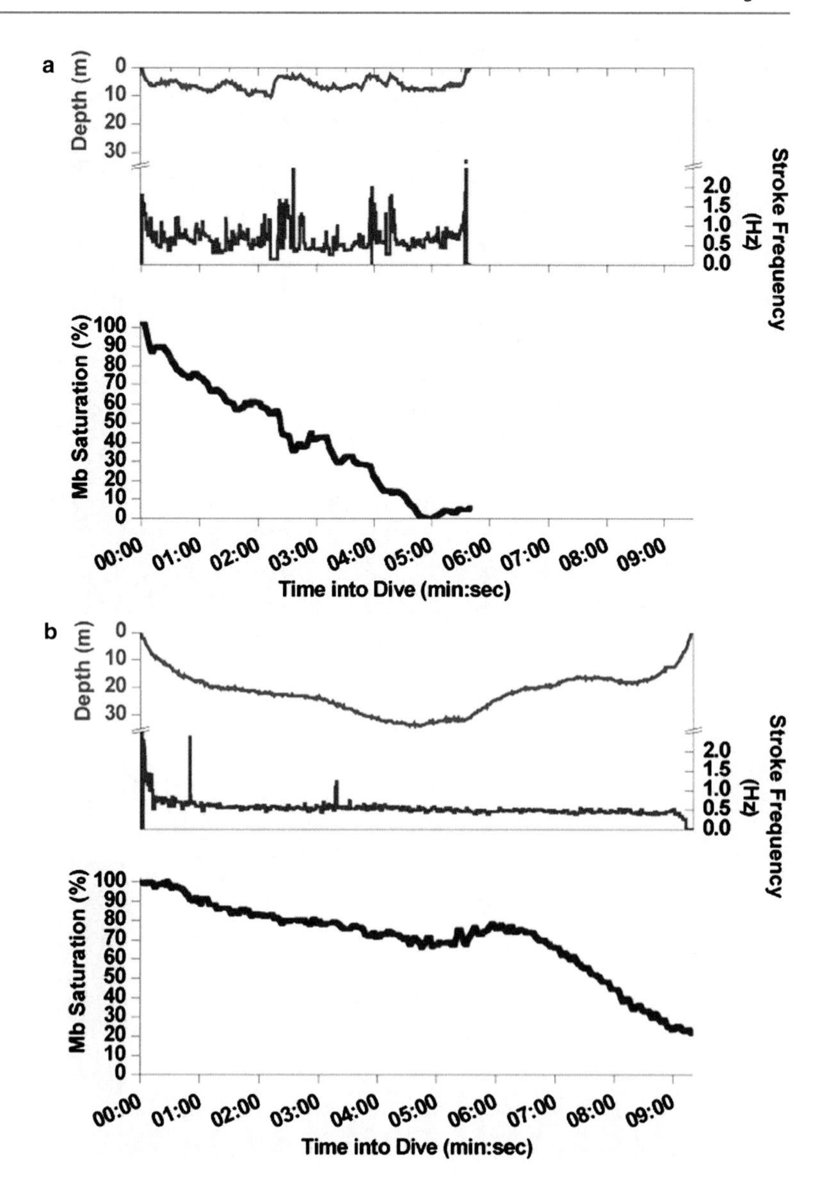

Fig. 7.5 Mb-O_2 saturation, dive depth, and stroke frequency for a 5.6-min type A dive (**a**) and a 9.3-min type B dive (**b**). By the end of the type A dive, approximately 5 min, Mb saturation is near 0%. At 5 min into the type B dive, Mb saturation has leveled off at 70%, and by the end of the type B dive Mb saturation is about 20% (From Williams et al. [2])

and supported the hypothesis that depletion of the muscle O_2 store and subsequent glycolysis was the source of the post-dive blood lactate increase in dives beyond the ADL [2]. Dives with this pattern (type A) were near or less than the ADL in duration.

Since this depletion pattern suggested no muscle blood flow and, based on prior research, lactate does not accumulate until dives beyond the ADL [10], the mean muscle O_2 consumption rate was determined. Using the mean O_2 desaturation rate from dives with a type A depletion pattern, the measured Mb concentration of 6.4 g Mb 100 g^{-1} muscle, and the known O_2 binding capacity of Mb of 1.34 ml O_2 g^{-1} Mb, mean muscle O_2 consumption was calculated to be 12.4 ml O_2 kg^{-1} muscle min^{-1} [2]. Muscle O_2 consumption (12.4 ml O_2 kg^{-1} muscle min^{-1}) during dives was only two to four times higher than resting muscle metabolic rates in dogs, seals, or humans [23, 34–38] and five to ten times lower than canine gastrocnemius stimulated at 0.25–0.5 Hz [39], demonstrating the efficiency of emperor penguin diving.

A second pattern of muscle O_2 depletion was also observed. In this pattern (type B), a much slower depletion rate occurred at the beginning of the dive, with a flat depletion rate often observed during the middle of these dives (Fig. 7.5b). In dives with a type B depletion pattern, which included dives from 2 to 12 min, stroke frequency was nearly constant, meaning muscle workload was not reduced. The slow depletion rate suggested muscle blood flow occurred during these dives (Fig. 7.5b). In Fig. 7.5b, between 4 and 6 min into the dive, although a virtual plateau in muscle O_2 depletion occurred, stroke frequency did decrease; in fact, stroke frequency was largely constant during most of the dive. Thus, as muscle continued to work during this plateau, maintenance of some muscle blood flow was indicated. Finally, at the end of the dive in Fig. 7.5b, the rate of muscle O_2 depletion was similar to the pattern observed in Figure 7.5a, which suggested that at the end of the dive, blood flow is cut off from the muscle. At the end of long dives with this second pattern, the Mb-O_2 store was often depleted. Thus, despite two different patterns, both were supportive of the hypothesis that the muscle O_2 store was the source of lactate. However, with the second depletion pattern, it appeared that penguins were able to delay the onset of lactate during these long dives. Researchers were unable to correlate any specific diving parameter (e.g., depth or duration) with a particular O_2 depletion pattern. This study showed for the first time that the muscle O_2 store is completely depleted in penguins during some voluntary dives and that penguins have at least two different muscle O_2 depletion patterns (Fig. 7.5) [2].

7.4.3 Muscle O_2 Depletion in Trained Harbor Seals

Jöbsis and colleagues used a Niroscope (Vander Corp., Durham, North Carolina, USA) to measure muscle O_2 depletion in trained and naïve harbor seals [3]. The Niroscope was a commercial device originally developed for monitoring tissue oxygenation in medical research and most commonly used for monitoring brain and muscle oxygenation (Patent #6594513). The Niroscope measures relative changes in Hb-oxygenation or trends for combined Mb-O_2 and Hb-O_2, combined deoxy-Mb (Mb) and deoxy-Hb (Hb), and total Hb (tHb), but does not provide absolute values. The relative changes in tissue oxygenation were reported in Vander Units (vd), a measure of relative change in Hb-O_2 + Mb-O_2, Hb + Mb, and tHb in an unknown volume of tissue [3]. The Niroscope probe was modified for the harbor seal study by replacing the surface probes with a fiber optic probe with four emitters at the probe tip and eight detectors set back 0.5 cm from the tip [3].

In this study, to simulate dives, seals wore a helmet that filled with water for a specific time period [3]. Seals were considered naïve when they experienced the water-filled mask for the first time. After a 1-week habituation to the mask and repeated dive simulations, the animals were considered trained [3]. Measurements of muscle O_2 depletion rates (using combined Hb-O_2 + Mb-O_2), heart rate, and muscle blood flow were compared between naïve and trained harbor seals (Fig. 7.6) [3].

This study demonstrated the difference in severity of three physiological responses between trained and forced (naïve) dives. However, the relationship is not a simple one-to-one relationship. Compared to trained seals, naïve seals experienced heart rates and muscle blood flows that were 50% and 33% of the respective values in trained seals (Fig. 7.6) [3]. In addition, naïve seals had 24% faster Hb-O_2 + Mb-O_2 depletion rates than in trained seals (Fig. 7.6). The faster "muscle O_2" depletion rate in naïve seals is consistent with more complete isolation of the muscle Mb-O_s store. However, describing muscle O_2 through interpretation of the NIRS signals is again complicated by the unknown contribution of Hb. In addition to the difference between naïve and trained dives, Fig. 7.6 also illustrates the response to a spontaneous breath hold. These results reinforce the severity in response observed during forced dives.

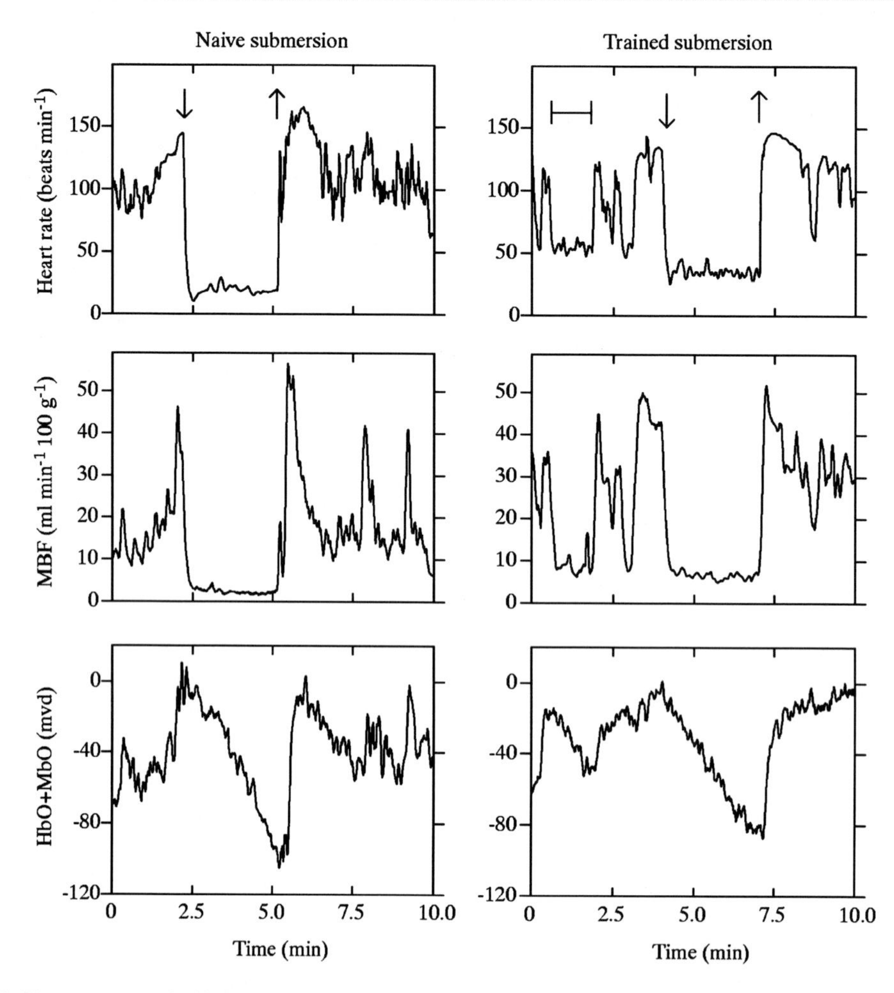

Fig. 7.6 Heart rate, muscle blood flow, and muscle O_2 ($Hb\text{-}O_2$ + $Mb\text{-}O_2$) of a seal during 3-min naïve and trained submersions. The start and end of submersions are indicated by *arrows*. The *line with brackets* indicates a 1.2-min spontaneous apnea before the trained submersion. Heart rate and muscle blood flow were much lower during the naïve submersion than the trained submersion and spontaneous apnea. Muscle O_2 depletion was also more severe in the naïve submersion (Reprinted with permission from Jöbsis et al. [3]. Copyright © 2011, The Company of Biologists)

7.5 Other Considerations in Instrument Design

In applying NIRS principles to the instruments described above, the use of such instruments on diving animals requires additional considerations. Motion artifact presents a particular challenge. Measuring $Mb\text{-}O_2$ saturation of the locomotory muscle during swimming necessarily requires that the muscle will be in constant motion as the animal strokes throughout the dive.

7.5.1 Probe Placement

In order to reduce movement of the probes, they were sutured to the surface of the muscle in the first two studies [1, 2]. Minimizing probe movement during the experiment was important for several reasons. First, movement of the probe would cause significant noise in NIRS readings. Second, the area which received the transmitted NIR light would change if the probe moved separately from the muscle during dives, causing inaccurate readings. Finally, movement of the probe over the surface of the muscle could cause trauma to the muscle (e.g., hematoma), also potentially obscuring NIRS readings.

7.5.2 Sampling Rate

Even with the probe sutured directly onto the pectoral muscle, movement artifact still occurred due to muscle movement in the emperor penguin study [2]. In order to account for predicted movement artifact due to the penguin stroke rate of 0.75 Hz [40], a 50-Hz sampling rate was used [2]. Thus, while there was movement artifact, the high sampling rate allowed for a good baseline between strokes. It does not appear that significant movement artifact occurred during the Weddell seal study, but this may be due to the choice of muscle, the latissimus dorsi [1].

7.6 Summary

This chapter has presented an important, if not unusual, application of NIRS principles that has increased our understanding of diving in marine animals. Using a simple application of NIRS principles, researchers have developed small, submersible NIRS recorders that can be used in the field with wild animals. This type of application is not available with other techniques, such as NMR.

The use of NIRS recorders has allowed for important discoveries on the use of the muscle O_2 store in diving mammals and birds, including potential differences between how the muscle O_2 store is used between two species of diving animals and the potential plasticity of the vascular response during diving.

Acknowledgments We would like to acknowledge funding support from NSF grant 0944220. CLW was supported by an NIH-NIAMS grant (T32AR047752 to V. Caiozzo) for the UC Irvine Multidisciplinary Exercise Sciences Training Program.

Problem

7.1. What is the ratio of Mb to Hb in a gram of the locomotory muscle of a seal versus that of a dog? (see [41–45]).

Further Reading

Guyton GP, Stanek KS, Schneider RC, Hochachka PW,Hurford WE, Zapol DG, Liggins GC, Zapol WM (1995) Myoglobin saturation in free-diving Weddell seals. J Appl Physiol 79:1148–1135

Jöbsis PD, Ponganis PJ, Kooyman GL (2001) Effects of training on forced submersion responses in harbor seals. J Exp Biol 204:3877–3885

Jöbsis PD (1998) Muscle oxygenation and blood flow during submersion in ducks (*Anas platyrhynchos*) and seals (*Phoca vitulina*). Ph.D. dissertation, University of California, San Diego, 142pp

Williams CL, Meir JU, Ponganis PJ (2011) What triggers the aerobic dive limit? Patterns of muscle oxygen depletion during dives of emperor penguins. J Exp Biol 214:1802–1812

References

1. Guyton GP, Stanek KS, Schneider RC, Hochachka PW, Hurford WE, Zapol DG, Liggins GC, Zapol WM (1995) Myoglobin saturation in free-diving Weddell seals. J Appl Physiol 79:1148–1155

2. Williams CL, Meir JU, Ponganis PJ (2011) What triggers the aerobic dive limit? Patterns of muscle oxygen depletion during dives of emperor penguins. J Exp Biol 214:1802–1812
3. Jöbsis PD, Ponganis PJ, Kooyman GL (2001) Effects of training on forced submersion responses in harbor seals. J Exp Biol 204:3877–3885
4. Butler PJ, Jones DR (1997) The physiology of diving of birds and mammals. Physiol Rev 77:837–899
5. Ponganis PJ (2011) Diving mammals. Compr Physiol 1:517–535. doi:10.1002/cphy.c091003
6. Burns JM, Skomp N, Bishop N, Lestyk K, Hammill M (2010) Development of aerobic and anaerobic metabolism in cardiac and skeletal muscles from harp and hooded seals. J Exp Biol 213:740–748
7. Scholander PF (1940) Experimental investigations on the respiratory function in diving mammals and birds. Hval Skrift 22:1–131
8. Scholander PF, Irving L, Grinnell SW (1942) Aerobic and anaerobic changes in seal muscle during diving. J Biol Chem 142:431–440
9. Kooyman GL, Wahrenbrock EA, Castellini MA, Davis RW, Sinnett EE (1980) Aerobic and anaerobic metabolism during voluntary diving in Weddell seals *Leptonychotes-weddelli* evidence of preferred pathways from blood chemistry and behavior. J Comp Physiol B 138:335–346
10. Ponganis PJ, Kooyman GL, Starke LN, Kooyman CA, Kooyman TG (1997) Post-dive blood lactate concentrations in emperor penguins, *Aptenodytes forsteri*. J Exp Biol 200:1623–1626
11. Ponganis PJ, Stockard TK, Meir JU, Williams CL, Ponganis KV, Howard R (2009) O_2 store management in diving emperor penguins. J Exp Biol 212:217–224
12. Elsner R (1965) Heart rate response in forced versus trained experimental dives of pinnipeds. Hval Skrift 48:24–29
13. Kooyman GL, Campbell WB (1973) Heart rate in freely diving Weddell seals, *Leptonychotes weddelli*. Comp Biochem Physiol A Physiol 43:31–36
14. Hill RD, Schneider RC, Liggins GC, Schuette AH, Elliott RL, Guppy M, Hochachka PW, Qvist J, Falke KJ, Zapol WM (1987) Heart rate and body temperature during free diving of Weddell seals. Am J Physiol 253:R344–R351
15. Fedak MA, Pullen MR, Kanwisher J (1988) Circulatory responses of seals to periodic breathing: heart rate and breathing during exercise and diving in the laboratory and open sea. Can J Zool 66:53–60
16. Thompson D, Fedak MA (1993) Cardiac responses of grey seals during diving at sea. J Exp Biol 174:139–164
17. Andrews RD, Jones DR, Williams JD, Thorson PH, Oliver GW, Costa DP, Le Boeuf BJ (1997) Heart rates of northern elephant seals diving at sea and resting on the beach. J Exp Biol 200:2083–2095
18. Ponganis PJ, Kooyman GL, Winter LM, Starke LN (1997) Heart rate and plasma lactate responses during submerged swimming and trained diving in California sea lions, *Zalophus californianus*. J Comp Physiol B 167:9–16
19. Ponganis PJ, Kooyman GL, Baranov EA, Thorson PH, Stewart BS (1997) The aerobic submersion limit of Baikal seals, *Phoca sibirica*. Can J Zool 75:1323–1327
20. Meir JU, Stockard TK, Williams CL, Ponganis KV, Ponganis PJ (2008) Heart rate regulation and extreme bradycardia in diving emperor penguins. J Exp Biol 211:1169–1179
21. Hindell MA, Lea MA (1998) Heart rate, swimming speed, and estimated oxygen consumption of a free-ranging southern elephant seal. Physiol Zool 71:74–84
22. Ponganis PJ, Kreutzer U, Sailasuta N, Knower T, Hurd R, Jue T (2002) Detection of myoglobin desaturation in *Mirounga angustirostris* during apnea. Am J Physiol Regul Integr Comp Physiol 282:R267–R272
23. Ponganis PJ, Kreutzer U, Stockard TK, Lin PC, Sailasuta N, Tran TK, Hurd R, Jue T (2008) Blood flow and metabolic regulation in seal muscle during apnea. J Exp Biol 211:3323–3332
24. Hassrick JL, Crocker DE, Teutschel NM, McDonald BI, Robinson PW, Simmons SE, Costa DP (2010) Condition and mass impact oxygen stores and dive duration in adult female northern elephant seals. J Exp Biol 213:585–592
25. Le Boeuf BJ, Morris PA, Blackwell SA, Crocker DE, Costa DP (1996) Diving behavior of juvenile northern elephant seals. Can J Zool 74:1632–1644
26. Le Boeuf BJ, Costa DP, Huntley AC, Feldkamp SD (1988) Continuous, deep diving in female northern seals, *Mirounga angustirostris*. Can J Zool 66:446–458
27. Blackwell SB, LeBoeuf BJ (1993) Developmental aspects of sleep apnea in northern elephant seals, *Mirounga angustirostris*. J Zool (Lond) 2331:437–447
28. Jöbsis FF (1977) Noninvasive, infrared monitoring of cerebral and myocardial oxygen sufficiency and circulatory parameters. Science 198:1264–1267
29. Mancini DM, Bolinger L, Li H, Kendrick K, Chance B, Wilson JR (1994) Validation of near-infrared spectroscopy in humans. J Appl Physiol 77:2740–2747
30. Tran TK, Sailasuta N, Kreutzer U, Hurd R, Chung Y, Mole P, Kuno S, Jue T (1999) Comparative analysis of NMR and NIRS measurements of intracellular PO_2 in human skeletal muscle. Am J Physiol Regul Integr Comp Physiol 276:R1682–R1690
31. Wilson JR, Mancini DM, McCully K, Ferraro N, Lanoce V, Chance B (1989) Noninvasive detection of skeletal muscle underperfusion with near-infrared spectroscopy in patients with heart failure. Circulation 80:1668–1674

32. Kanatous SB, DiMichele LV, Cowan DF, Davis RW (1999) High aerobic capacities in the skeletal muscles of pinnipeds: adaptations to diving hypoxia. J Appl Physiol 86:1247–1256
33. Howell AB (1930) Aquatic mammals. Dover, New York
34. Duran WN, Renkin EM (1974) Oxygen-consumption and blood-flow in resting mammalian skeletal muscle. Am J Physiol 226:173–177
35. Hogan MC, Nioka S, Brechue WF, Chance B (1992) A ^{31}P-NMR study of tissue respiration in working dog muscle during reduced O_2 delivery conditions. J Appl Physiol 73:1662–1670
36. Hogan MC, Kurdak SS, Arthur PG (1996) Effect of gradual reduction in O_2 delivery on intracellular homeostasis in contracting skeletal muscle. J Appl Physiol 80:1313–1321
37. Piiper J, Pendergast DR, Marconi C, Meyer M, Heisler N, Cerretelli P (1985) Blood flow distribution in dog gastrocnemius muscle at rest and during stimulation. J Appl Physiol 58:2068–2074
38. Mizuno M, Kimura Y, Iwakawa T, Oda K, Ishii K, Ishiwata K, Nakamura Y, Muraoka I (2003) Regional differences in blood flow and oxygen consumption in resting muscle and their relationship during recovery from exhaustive exercise. J Appl Physiol 95:2204–2210
39. Hogan MC, Gladden LB, Grassi B, Stary CM, Samaja M (1998) Bioenergetics of contracting skeletal muscle after partial reduction of blood flow. J Appl Physiol 84:1882–1888
40. van Dam RP, Ponganis PJ, Ponganis KV, Levenson DH, Marshall G (2002) Stroke frequencies of emperor penguins diving under sea ice. J Exp Biol 205:3769–3774
41. Kanatous SB, Elsner R, Mathieu-Costello O (2001) Muscle capillary supply in harbor seals. J Appl Physiol 90:1919–1926
42. Karas RH, Taylor CR, Rosler K, Hoppeler H (1987) Adaptive variation in the mammalian respiratory system in relation to energetic demand, V: limits to oxygen transport by the circulation. Respir Physiol 69:65–79
43. Kayar SR, Hoppeler H, Jones JH, Longworth K, Armstrong RB, Laughlin MH, Lindstedt SL, Bicudo JE, Groebe K, Taylor CR et al (1994) Capillary blood transit time in muscles in relation to body size and aerobic capacity. J Exp Biol 194:69–81
44. Polasek LK, Dickson KA, Davis RW (2006) Metabolic indicators in the skeletal muscles of harbor seals *Phoca vitulina*. Am J Physiol Regul Integr Comp Physiol 290:1720–1727
45. Willford DC, Gray AT, Hempleman SC, Davis RC, Hill EP (1990) Temperature and the oxygen–hemoglobin dissociation curve of the harbor seal, *Phoca vitulina*. Respir Physiol 79:137–144

Noninvasive NMR and NIRS Measurement of Vascular and Intracellular Oxygenation In Vivo

8

Youngran Chung and Thomas Jue

8.1 Introduction

In the canonical biochemical view, respiration requires a coordinate regulation of oxygen transport, carbon unit flux, electron flow, proton pumping, and ADP translocation. Certainly in-vitro experiments have supported the paradigm, cast in terms of either a kinetic or thermodynamic model [1]. Yet, in blood-perfused tissue the specific regulatory mechanisms still remain unclear [2, 3]. In contrast to isolated mitochondria or cells, many overlapping reactions intervene to regulate cellular bioenergetics [4].

In particular, muscle alters dramatically its bioenergetics during sudden transition from rest to work. As such, muscle can serve as a convenient model to study how the cell regulates dynamically its metabolism and oxygen consumption (VO_2): upon initiation of contraction, the immediate rise in cellular energy demand triggers a rapid adaptation in the ventilatory, cardiovascular, and peripheral blood flow response to enhance O_2 flux. How the cell mobilizes its metabolic resources, increases its metabolic rates, and adjusts O_2/nutrient flux pose central questions in the theory of respiratory control. Although some researchers downplay a significant role for a metabolic regulation of VO_2 and point to a tight match between O_2 supply throughout contraction, others have contended that metabolism must modulate VO_2 demand and blood flow [5].

From the vantage of cellular bioenergetics, experiments have measured metabolic fluxes and have analyzed the results to discriminate kinetic versus thermodynamic control [1]. Specifically, in the kinetic model, a substrate such as ADP limits the reaction rate [2]. Since muscle contraction consumes ATP, the model predicts that the product of ATP hydrolysis, ADP, regulates the rate of respiration to match the rate of energy utilization. In contrast, the thermodynamic model posits a balanced change in key metabolite levels, which act conjointly to regulate respiration.

However, in-vivo NMR myocardium studies have failed to confirm these simple ideas. Elevating VO_2 does not elicit any rise in ADP, as derived from the ^{31}P signals of PCr and ATP [6–8]. This observation does not align precisely with the prediction of either the kinetic or thermodynamic models. In contrast, skeletal muscle shows a drop in PCr level during contraction and an ADP-dependent VO_2 [9]. But the ADP-dependent rise in VO_2 does not appear at the initial phase of muscle contraction [10].

Y. Chung, M.D., Ph.D. (✉) • T. Jue, Ph.D. (✉)
Biochemistry and Molecular Medicine, University of California Davis, 95616-8635 Davis, CA, USA
e-mail: yrchung@ucdavis.edu; tjue@ucdavis.edu

T. Jue and K. Masuda (eds.), *Application of Near Infrared Spectroscopy in Biomedicine*,
Handbook of Modern Biophysics 4, DOI 10.1007/978-1-4614-6252-1_8,

Resolving these discordant observations requires an accurate assessment of intracellular VO_2 and the O_2 supply. Because whole body VO_2 and arterial venous oxygen difference measurements with a mean end capillary blood flow have limited accuracy in determining the dynamic change in VO_2 in localized myocytes, researchers have explored noninvasive in-vivo techniques, such as near-infrared spectroscopy (NIRS) and nuclear magnetic resonance (NMR) [11, 12]. Certainly, air-breathing animals require O_2 to survive. But how the cell modulates its O_2 utilization to meet a physiological range of energy demand remains an open question.

8.2 Microelectrode Measurement of Vascular and Intracellular Oxygenation

A simple, direct approach to measuring tissue O_2 utilizes microelectrodes. These electrodes are usually comprised of a platinum wire with a recessed cathode tip of ~1 μm. They respond to changes in O_2 within 25 ms. For in-vivo application, however, they have potential limitations. They can inflict tissue damage, have low spatial and temporal resolution, might consume oxygen at the cathode to alter the tissue environment, and can introduce artifacts arising from fluid convection or by variation in oxygen transport [13–17]. Moreover, any rigorous interpretation of the in-vivo experimental data requires sophisticated mathematical models to guide the analysis [18, 19]. These concerns limit the use of O_2 electrodes to study human subjects during dynamic movement. Nevertheless, O_2 microelectrode studies in animal models have revealed unique physiological insights: tissue has a lower pO_2 than venous blood, which implies the presence of heterogeneous distribution of oxygen in the microcirculation [20].

8.3 Spectroscopic Approaches

Instead of using microelectrodes to measure O_2 in tissue, other techniques have used the signals from oxygen-binding heme proteins: myoglobin (Mb) and hemoglobin (Hb). Hb exists exclusively in the vasculature, while Mb localizes in the cytoplasm of myocytes. Upon binding O_2, the electronic configuration of the heme Fe^{2+} converts from a paramagnetic to a diamagnetic state. Because the electronic configuration and protein structure of the oxygenated and deoxygenated heme protein differ, the associated spectroscopic signatures can characterize the different oxygenation states. Given the relative ratio of deoxy-Mb or deoxy-Hb (dMb or dHb) and oxy-Mb or Hb (MbO_2 or HbO_2) signals, a calculation using the in-vitro association (K_a) or dissociation constants (K_d) of O_2 binding to myoglobin and hemoglobin leads to a value for the partial pressure of oxygen (pO_2) in the vascular and/or cellular space.

Association of O_2 from Mb follows the expressions

$$[\mathrm{Mb}] + [\mathrm{O_2}] \Leftrightarrow [\mathrm{MbO_2}],$$

$$\mathrm{K} = \frac{[\mathrm{MbO_2}]}{[\mathrm{Mb}][\mathrm{O_2}]},$$

where $[MbO_2]$ is the concentration of oxy-Mb, [Mb] is the concentration of deoxy-Mb, $[O_2]$ is free O_2, and K is the dissociation constant. Since Mb binds a single O_2, the O_2 saturation of Mb becomes

$$\mathrm{S}_{\mathrm{MbO_2}} = \frac{[\mathrm{MbO_2}]}{[\mathrm{Mb}] + [\mathrm{MbO_2}]} = \frac{\mathrm{K}[\mathrm{Mb}][\mathrm{O_2}]}{[\mathrm{Mb}] + \mathrm{K}[\mathrm{Mb}][\mathrm{O_2}]},$$

where S_{MbO_2} is the fraction of Mb saturated with O_2. Converting the equation in terms of pO_2 and P_{50} (the partial pressure of oxygen and partial pressure of oxygen that saturates half the Mb, respectively) transforms the equation into

$$S_{MbO_2} = \frac{K[O_2]}{1 + K[O_2]} = \frac{[O_2]}{\frac{1}{K} + [O_2]} = \frac{pO_2}{P_{50} + pO_2},$$

For Hb, the equation includes exponent n (the Hill coefficient) to account for the additional heme O_2-binding sites in tetrameric Hb:

$$S_{HbO_2} = \frac{K[O_2]^n}{1 + K[O_2]^n} = \frac{[O_2]^n}{\frac{1}{K} + [O_2]^n} = \frac{pO_2{}^n}{P_{50} + pO_2{}^n}.$$

Many studies have determined the O_2 binding curve of Mb and HbA (normal human adult Hb), characterized by its O_2 affinity (specified usually in terms of its P_{50}) and Hill coefficient. Monomeric Mb has a P_{50} of 2.3 mmHg at 35°C, with $n = 1$. HbA has a P_{50} of about 26 mmHg and can take Hill coefficient values from 1 to 4. For HbA, experiments have determined a Hill coefficient around 2.8 and have interpreted it as an index of protein cooperativity [21].

For in-vivo application, blood volume alteration can confound the analysis. Because muscle contraction elicits vasodilatation and vasoconstriction, blood volume can change and modulate overall spectrum intensity without shifting the ratio of the oxy- and deoxyhemoglobin signals. Monitoring overall spectral intensity changes in the oxy- and deoxy-Hb signal provides an index of relative blood volume change. Alternatively, following the isosbestic points in the transition from oxy- to deoxy-Hb can also help delineate blood volume from oxygenation changes (vide infra). The ratio of deoxy-Hb/HbO_2 reflects the oxygenation state.

8.4 Hb Absorbance of Visible Light

Figure 8.1 displays the typical spectra from both deoxy-Hb (dHb) and HbO_2 in part of the visible spectral region 450–650 nm. For HbA, the HbO_2 spectrum shows a distinct Soret band (γ) at 415 nm (not shown). Two additional peaks (α and β bands) appear at 577 and 541 nm. As oxygen levels decrease, the Soret γ band shifts from 415 toward 430 nm, while the α and β bands gradually disappear.

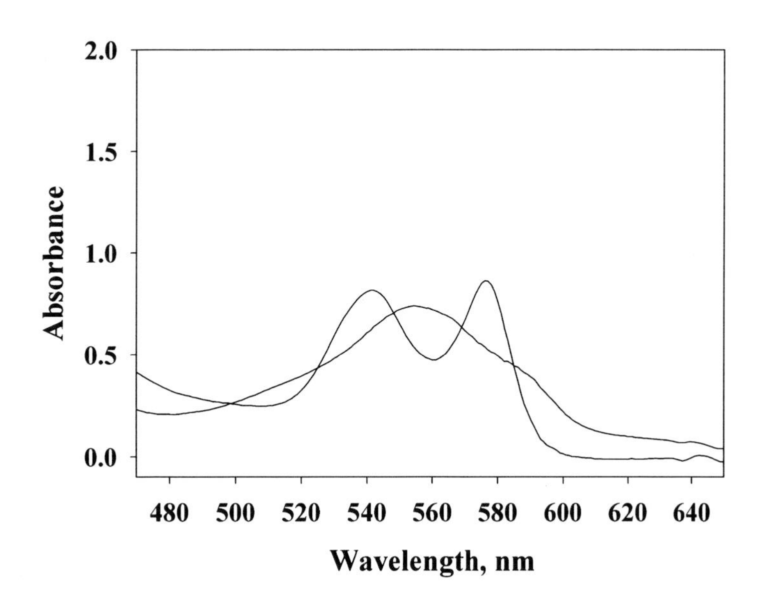

Fig. 8.1 **Visible spectra of HbO_2 and deoxy-Hb.** HbO_2 has peak maxima at 541 and 577 nm, and deoxy-Hb has a signal maximum at 555 nm

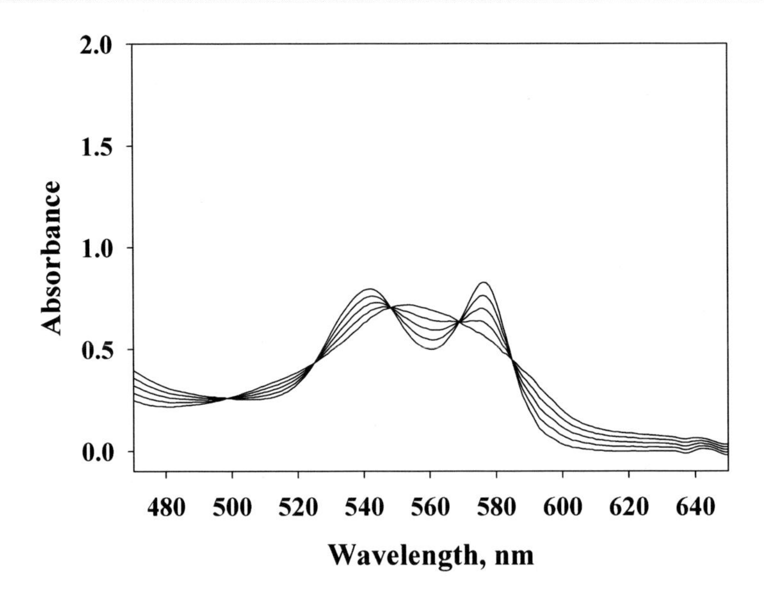

Fig. 8.2 **Spectra simulating the Hb signals at varying HbO_2/deoxy-Hb molar ratios: 1:9, 3:7, 5:5, 7:3, 9:1**. The spectra reflect the varying levels of blood oxygenation. Isosbestic points occur at 498, 525, 548, 568, and 585 nm. Signal intensity at the HbO_2 and deoxy-Hb maxima vary, but the isosbestic points remain invariant

A new peak from dHb emerges at 555 nm. The HbO_2 (α and β bands), the deoxy-Hb signal, the HbO_2 signal, and the association or dissociation constant (K_a or K_d) lead to a determination of pO_2.

Figure 8.2 models a bank of HbO_2 and dHb spectra during transition from the deoxygenated to the oxygenated state. The signal at 555 nm decreases as the signals at 541 and 577 nm increase, as the HbO_2/deoxy-Hb molar ratios increase: 1:9, 3:7, 5:5, 7:3, 9:1. Even though the HbO_2 and deoxy-Hb signal intensities change with oxygenation, the spectra intersection points or isosbestic points at 498, 525, 548, 568, and 585 nm remain invariant. However, if the overall Hb signal concentration or volume increase, as observed during vasodilation, the isosbestic points will rise. Figure 8.3 shows a factor of twofold increase in blood at HbO_2/deoxy-Hb molar ratios 1:9, 3:7, 5:5, 7:3, and 9:1, respectively. Compared with Fig. 8.1, the overall spectral intensity has increased by a factor of two, as reflected in the isosbestic points, but the signal intensity ratio of HbO_2 and deoxy-Hb remains the same. The isosbestic points then provide a way to normalize Hb saturation with respect to the total Hb pool.

8.5 Mb Absorbance of Visible Light

Because both Mb and Hb have Fe-protoporphyrin IX as the prosthetic group and share protein structural features, such as a proximal and a distal histidine near the Fe center, the porphryin $\pi - \pi^*$ transition (a_{1u}, a_{2u} to e_g (π^*)) gives rise to a similar Soret band as well as the α and β bands (sometimes referred to as the Q_o and Q_v bands) [22]. For MbO_2, the α and β bands appear at 580 and 542 nm. For HbO_2, the corresponding bands also appear around 580 and 542 nm and exhibit similar extinction coefficients. Separating the MbO_2 and HbO_2 signals in the visible spectra of blood-perfused muscle then poses a formidable challenge. In contrast, brain tissue has no Mb to interfere with Hb detection. Nevertheless, the Hb and Mb signals reflect the distinct vascular versus intracellular oxygenation.

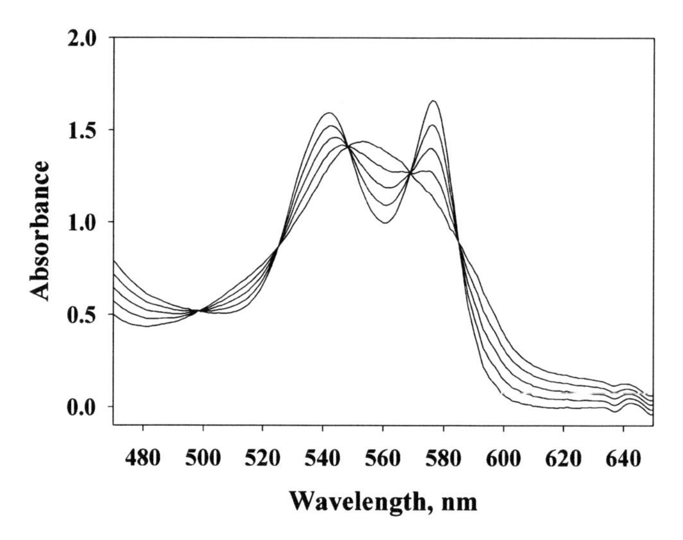

Fig. 8.3 **Spectra simulating blood volume and Hb oxygenation changes**. The overall HbO_2 and deoxy-Hb signals will increase with blood volume. The visible spectra reflect a twofold increase in total blood volume at different HbO_2/deoxy-Hb molar ratios: 1:9, 3:7, 5:5, 7:3, 9:1, respectively. In comparison to Fig. 8.2, the isosbestic points occur at the same wavelengths, but the intensity has risen by a factor of 2. The ratios of HbO_2/deoxy-Hb signal intensities, however, remain the same

For in-vitro assay, many studies have employed a Reynafarje algorithm to discriminate Mb from Hb, which is predicated on the specific spectral features of MbCO and HbCO [23]. Because HbCO displays a β-to-α intensity ratio of 1, while MbCO shows a ratio of about 0.8, the signal intensity difference at 568 and 538 nm for a mixture of MbCO and HbCO can factor out the Mb concentration in samples extracted from blood-perfused tissue:

$$A_{538} - A_{568} = (\varepsilon_{538\mathrm{HbCO}} - \varepsilon_{568\mathrm{HbCO}})C_{\mathrm{HbCO}} + (\varepsilon_{538\mathrm{MbCO}} - \varepsilon_{568\mathrm{MbCO}})C_{\mathrm{MbCO}}.$$

Since $\varepsilon_{538\mathrm{HbCO}} = \varepsilon_{568\mathrm{HbCO}}$,

$$C_{\mathrm{MbCO}} = \frac{A_{538} - A_{568}}{\varepsilon_{538\mathrm{MbCO}} - \varepsilon_{568\mathrm{MbCO}}} = \frac{A_{538} - A_{568}}{14.7 \times 10^3 - 11.8 \times 10^3} = (A_{538} - A_{568}) \times 3.45 \times 10^{-4}\ \mathrm{M}.$$

Despite its simple elegance, an unmindful application of the Reynafarje algorithm without an appropriate calibration and baseline adjustment can lead to an erroneous determination of the Mb concentration and a consequent misinterpretation of Mb function [24]. Moreover, the $\varepsilon_{538\mathrm{HbCO}} = \varepsilon_{568\mathrm{HbCO}}$ condition doesn't always hold; the values for $\varepsilon_{538\mathrm{MbCO}} - \varepsilon_{568\mathrm{MbCO}}$ vary with different species, and small changes in the spectral baseline will distort the analysis.

8.6 Near-Infrared Versus Visible Light Scattering

Using visible light to follow vascular or intracellular changes in pO_2, however, appears limited by its shallow (millimeter) penetration depth. Light propagation attenuates (μ_t) by scattering (μ_s) and absorbance (μ_a) interactions with molecules in the medium, as expressed by

$$\mu_t = \mu_s + \mu_a,$$

where $1/\mu_t$ defines the mean free path, the mean distance a molecule travels, before it encounters any scattering or absorbance interaction.

Water dominates in most biological tissue and represents about 70–80% of overall content (~55 M). It represents the major source for scattering

Water also has a broad absorbance with a peak around 1,000 nm. As a result, the water signal contributes significantly to the NIRS spectra from 750 to 1,400 nm. The scattering coefficient (μ_s) also has a specific dependence upon wavelength:

$$\mu_s = A\lambda^{-P},$$

where A is the scattering amplitude and P is the scattering power. A reflects the scattering strength. P relates to scattering particle average size. For particles with diameter d, much smaller than the wavelength of light ($d \ll \lambda$), the Rayleigh scattering limit $P = 4$. P decreases with increasing particle size (Mie scattering). The ratio R of light penetration depths at wavelengths λ_1 and λ_2 ($\lambda_1 < \lambda_2$), at a specific P, and with the scattering coefficients, follows the equation

$$R = \frac{\mu_t^{\lambda_2}}{\mu_t^{\lambda_1}} = \frac{\mu_s^{\lambda_1} + C\mu_a^{\lambda_1}}{\left(\frac{\lambda_1}{\lambda_2}\right)^P \mu_s^{\lambda_1} + C\mu_a^{\lambda_2}}.$$

This equation indicates that scattering will decrease as wavelength increases. Penetration depth will increase correspondingly. Specifically, using NIRS instead of visible light extends the penetration depth from a millimeter to the centimeter range [25–27].

Indeed, with a typical source–detector separation of 3 cm, NIRS will detect signals from a banana-shaped volume centered approximately 1.5 cm below the surface [28]. A larger spacing between source and detector will sample signals from a deeper volume. For brain studies, spatial- and time-resolved NIRS experiments have ascribed 55% and 69% of the detected signal arising from cerebral tissue. Several reviews have discussed NIRS signal quantification, advantages/limitations, and biomedical applications [29–31].

8.7 Interference from Fat Layer

Hb also absorbs near-infrared light (750–1,400 nm) (Fig. 8.4). DHb has a well-defined absorbance maximum at 760 nm. In contrast, HbO_2 has a broad absorbance extending from 750 to 1,000 nm. Many NIRS instruments use the absorbance at 850 nm to denote the HbO_2 concentration, partly because it skirts the lipid signal centered at 940 nm and the intense water signal at 1,000 nm (Fig. 8.4).

Lipid, however, has a near-infrared absorbance. It interferes prominently in tissue with high ratios of lipid to Hb and with thick fat layers. Because subcutaneous tissue contains a high concentration of fat, the extent of interference also depends on penetration depth. With close optode spacing, the detected NIRS signal has a significant contribution from the shallow layers containing subcutaneous fat. As optode spacing increases to sample signals from deeper tissue, fat interference decreases. This variation in fat contribution based on optode spacing provides a basis for deconvoluting HbO_2 from the fat signal. Such an approach produces a linear intensity relationship with penetration depth using signals from optode separation at 7, 20, 30, and 40 mm. The 7-mm spaced optodes also help to minimize the skin effect. Monte Carlo simulations confirm the strategy to deconvolute the fat layer signal. With the fat correction algorithm, the coefficient of variation (standard deviation/mean) for the NIRS HbO_2 signal in exercising muscle decreases from 40–50% to 11–21% [32, 33].

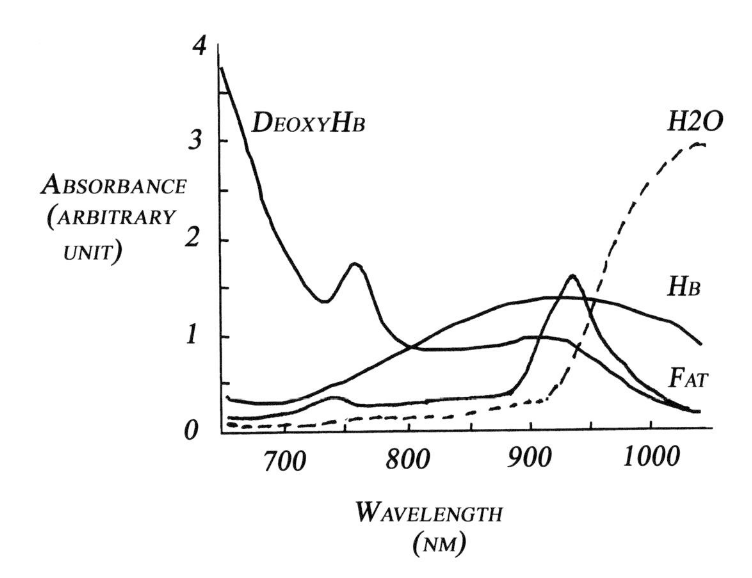

Fig. 8.4 Representative 650–1,050 nm traces denoting the contribution from HbO_2, deoxy-Hb, fat, and water

8.8 Overlapping Mb and Hb Signals

As in the case of the visible spectrum, the Mb and Hb NIRS signals overlap. Some researchers have ascribed no significant role for Mb to supply O_2 during muscle contraction. Because Mb has an extremely high affinity for O_2, it cannot readily release its O_2 store. Hb must then supply all the O_2 from the very onset of contraction [34–36]. Others have assumed that NIRS monitors predominantly Hb oxygen saturation and desaturation kinetics. As a consequence, the NIRS observed change (Δ[deoxy]) arises from the capillary blood flow adjustment, which supports a school of thought that envisions a change in the gradient from capillary O_2 to the muscle that precisely matches O_2 demand (VO_2) [37, 38]. Indeed, many NIRS experiments assume a dominant Hb contribution [29, 39]. NMR and modeling experiments, however, show that Mb does desaturate and can contribute significantly to the NIRS signal.

Even though the signals overlap, Mb and Hb do exhibit slight spectral differences. Under ideal conditions, a second-derivative transformation reveals the spectral differences and provides the basis for a wavelength shift analysis to distinguish Mb from Hb [40]. Such an approach shows that Hb contributes 20% to the NIRS signal from human muscle, but 87% from mouse muscle.

NIRS measurements during contraction, however, introduce prominent motional artifacts, which would compromise the accuracy and precision of the wavelength shift analysis approach. Many researchers simply filter or smooth out these noise artifacts during data analysis. Others have culled these cyclic NIRS signals to determine vascular oxygenation during muscle contraction [41].

Most commercial NIRS instruments utilize the diffusion equation and the absorption coefficients (μ_a) to derive muscle oxygen saturation (SmO_2). The solution to this equation assumes a particular geometry, such as a homogenous semiinfinite half-space geometry, and an optode configuration, and requires experimental input parameters to determine the absorption coefficients (μ_a). The μ_a measured at multiple wavelengths then leads to tissue oxygen saturation via the Beer-Lambert law. However, the model assumption of a homogenous medium quickly departs from experiment reality since the fat and skin layers introduce inhomogeneities in muscle. Muscle, skin, and fat layers will attenuate light.

8.9 NMR

^{1}H NMR can also monitor cellular and vascular oxygenation by following the signals of cytosolic myoglobin and blood hemoglobin [42]. A comparative analysis of NIRS and NMR data would clarify the relative contributions in the NIRS spectra.

Myoglobin is composed of approximately 153 amino acids and has a molecular weight of 17,500. It has an 85% helical structure and contains eight distinct segments (A–H). The iron in the prosthetic heme group coordinates oxygen with a binding affinity about ten times higher than that of Hb [21]. Despite the conservation of key residues, the primary sequence shows extensive variability across species [43]. How the primary, secondary, and tertiary structures modulate myoglobin's binding affinity for oxygen poses the central question in protein structure–function studies [44]. Mammalian hemoglobin also has a heme prosthetic group in each of its four subunits.

Under physiological conditions, the heme Fe exists predominantly in the +2 oxidation state. Ligated with oxygen, the heme Fe(II) electrons become paired ($S = 0$) or diamagnetic. Unligated with oxygen, the heme Fe(II) electrons become unpaired ($S = 2$) or paramagnetic [45]. The unpaired electrons in the paramagnetic state interact with the proton and produce a hyperfine shift in the NMR signal. Such a hyperfine interaction can originate from either a contact (through bond) or pseudocontact (through space) contribution [46, 47].

$$\left(\frac{\Delta H}{H}\right)_{\text{hyperfine}} = \left(\frac{\Delta H}{H}\right)_{\text{contact}} + \left(\frac{\Delta H}{H}\right)_{\text{pseudocontact}},$$

where H is the magnetic field, $(\Delta H/H)_{\text{hyperfine}}$ is the overall hyperfine interaction, $(\Delta H/H)_{\text{contact}}$ is the contact contribution, and $(\Delta H/H)_{\text{pseudocontact}}$ is the dipolar or pseudocontact contribution.

In the contact mechanism, the hyperfine shift depends upon an electron–nuclear coupling and the molecular orbital overlap:

$$\left(\frac{\Delta H}{H}_{\text{con}}\right) = -\mathrm{A}\frac{g\beta S(S+1)}{3\gamma kT},$$

where A is the electron-nuclear coupling constant, β is the Bohr magneton, γ is the magnetogyric ratio, g is the Lande constant, S is the spin quantum number, k is Boltzmann's constant, and T is the absolute temperature.

In the pseudocontact mechanism, the unpaired electron generates a dipole field:

$$\left(\frac{\Delta H}{H}\right)_{\text{pcon}} = \frac{1}{3R^3}\{(1-3\cos^2\theta)(X_{zz} - \frac{1}{2}(X_{xx} + X_{yy}) + \sin^2\theta\cos 2\Omega(X_{yy} - X_{xx})\},$$

where R is the internuclear distance, θ and Ω are angles in the molecular axis system, and X_{nn} ($n = x, y, z$) are diagonal elements of the magnetic susceptibility tensor.

The contact shift reflects the through bond, the molecular orbital interaction, while the pseudocontact shift reflects the through space, the dipole interaction. Separating the contributions can pose some technical challenges [48, 49]. The paramagnetism also gives rise to a bulk magnetic susceptibility, given by the Curie law:

$$X = \frac{Ng^2\beta^2 S(S+1)}{3kT} = \frac{C}{T},$$

where N is the number of spins, g is Lande's constant, β is the Bohr magneton, S is the spin quantum number, k is Boltzmann's constant, and T is the absolute temperature.

8.10 Assignment of the Proximal Histidyl $N_\delta H$ Signal

Model compound and mutant protein studies have led to specific NMR assignments. In particular, the model study of 2-methyl imidazole axially coordinated to tetraphenylporphyrin (TPP) led to assignment of the proximal histidyl $N_\delta H$ signal of Mb and Hb [50]. The exchangeable 2-methyl imidazole attached to the paramagnetic Fe^{2+} has its $N_\delta H$ signal appearing at ~80 ppm. In deuterated solvent, the signal disappears. Its hyperfine shift originates predominantly from electron delocalization via the contact shift mechanism. Magnetic anisotropy, associated with the dipolar shift mechanism, contributes only about 10%.

Indeed, ^1H NMR studies of deoxygenated Mb and Hb have detected exchangeable signals at ~80 ppm and have assigned them to the proximal histidyl $N_\delta H$ [51, 52]. Mb shows one signal. Because of the inequivalence of the α and β subunits, HbA exhibits two proximal histidyl $N_\delta H$ signals. At 25°C, HbA exhibits signals at 76 and 64 ppm, corresponding to the β and α subunits, respectively [53]. The corresponding signal for human myoglobin appears at 81 ppm [54]. Upon oxygenation, the signal disappears (see Fig. 8.5).

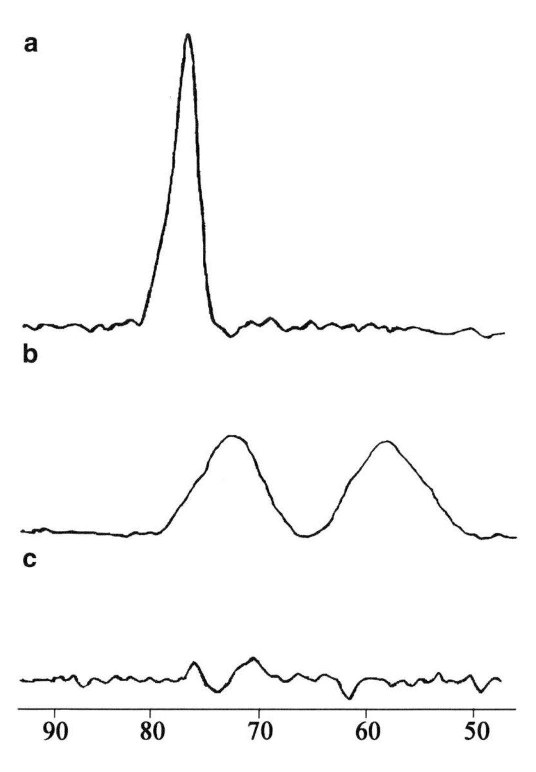

Fig. 8.5 **^{1}H NMR spectra in the region of 80 ppm.** (**a**) Human muscle tissue under resting condition. No detectable deoxy-Mb signal; (**b**) proximal histidyl $N_\delta H$ signals from α (64 ppm) and β (76 ppm) subunits of deoxy-Hb in erthryocte; (**c**) deoxy-Mb proximal histidyl $N_\delta H$ signal (80 ppm) from human gastrocnemius muscle after cuffing the blood flow

Although the proximal histidyl $N_\delta H$ proton can exchange with water, the exchange lifetime spans over a long NMR timescale. For deoxy-HbA the proximal histidyl NH exchange half-life occurs on the order of minutes to hours [55]. The α heme pocket has greater solvent accessibility than the β subunit.

8.11 Assignment of the Val E11 CH_3 Signal

The π delocalized electrons in the diamagnetic MbO_2 create a local magnetic field that interacts with local amino-acid residues in the heme pocket and shifts their resonance position. In particular, the Mb Val E11 γ1 CH_3 near the distal side of the heme experiences a significant ring current shift to −2.8 ppm [56]. The corresponding Hb α and β Val E11 γ1 CH_3 residues experience a similar shift to −2.39 ppm [57]. Because CO and O_2 bind to the heme Fe in different geometric configurations and will alter the local magnetic field, the Val E11 signals of MbCO and HbCO have altered resonance positions: MbCO (−2.40 ppm), HbCO-α (−1.70 ppm), and HbCO-β (−1.82 ppm) [53]. These upfield Val Ell signals decrease with decreasing oxygenation.

8.12 NMR and Measurement of Tissue Oxygenation

^{1}H NMR can detect Mb Val E11 and proximal histidyl $N_\delta H$ signals in cardiac and skeletal muscle in vivo. During hypoxia or ischemia, Mb releases O_2 in the myocytes, as reflected in the rising signal of deoxy-Mb proximal histidyl $N_\delta H$ or the decreasing signal of Val E11 [58–61]. The decreasing intracellular O_2 level implicates a widening O_2 gradient from the capillary to the cell to facilitate O_2 flux into the cell [59, 61, 62]. Whether the gradient continues to widen or plateaus with increasing energy demand remains moot. Nevertheless, Mb desaturates with a rapid kinetics, consistent with a transient mismatch of O_2 supply and demand [10].

Contrary to the viewpoint of some researchers, Mb can contribute significantly to the NIRS signal. Indeed, perfused hindquarter muscle shows a dynamic range of Mb and Hb deoxygenation during graded levels of exercise intensity under blood and buffer perfusion. The results indicate that the NIRS signal contains a major contribution from Mb. Mb rapidly desaturates at the onset of muscle contraction and the intracellular oxygen level decreases proportionately with increasing work, with no observable plateau [62].

Because NMR can detect the distinct Mb and Hb signals of Val E11 and the proximal histidyl $N_\delta H$, it can help shed light on the contribution of Mb versus Hb in the NIRS signal and the role of O_2 supply and utilization in respiratory regulation during contraction. In fact, the respiration difference during exercise in normal versus heart failure subjects does not arise from a deficient O_2 supply. Instead, it appears to originate from an inadequate O_2 utilization in mitochondria [63].

When blood flow to the human gastrocnemius muscle decreases, the NMR spectra clearly exhibit the deoxy-Mb signal at 78 ppm, which rises as oxygenation decreases. Upfield at 73 ppm a signal corresponding to the deoxy-Hb β-subunit histidyl F8 $N_d H$ appears and reaches a steady-state level [59]. Under the same experimental protocol, the kinetics of Mb desaturation matches the NIRS-observed decline in the composite MbO_2 and HbO_2 signal. Even after Mb has attained a steady-state level, the high-energy phosphate levels, as reflected in the ^{31}P spectra, remain unperturbed.

Because Hb and Mb desaturation kinetics depends critically on the cuffing and exercise protocols, comparative NIRS and NMR analysis has validity only when the data originate from a consistent set of experiments with well-defined cuffing protocols. Using the observation of a fast NIRS desaturation kinetics from one experiment and a slow NMR Mb desaturation kinetics from another to argue against a significant Mb contribution in the NIRS signal would not constitute a compelling argument [64]. Another combined ^{1}H-NMR and NIRS study using human and canine muscles also ascribes more than 50% of the NIRS signal to Mb [65]. A computer model indicates that the Mb contribution can represent 53–86% of the overall NIRS signal [66].

8.13 Mechanism of Respiratory Control

The kinetics of Mb and Hb desaturation during exercise at different workloads would yield invaluable insights into the canonical view of respiratory control, which posits a tight coordination of electron flow, proton pumping, O_2 supply/delivery, metabolic flux, and an ADP-dependent feedback control.

In-vitro experiments certainly support the standard paradigm, cast in either a kinetic or a thermodynamic model. In blood-perfused tissue, however, numerous reaction steps can intervene to regulate respiration [2]. The sudden energy demand during muscle contraction stimulates both cellular metabolism and an O_2 delivery. How the cell adapts to this surge in energy demand stands central to the question of respiratory control. Unfortunately, NIRS at present does not yield definitive insight, since it cannot easily distinguish intracellular from vascular O_2 or VO_2.

^{1}H NMR studies have already shown an approach to map the intracellular pO_2 with the Mb signals in vivo [58, 67, 68]. These NMR studies have reported fully saturated MbO_2 in resting muscle, implying that the resting myocyte pO_2 is well above an Mb P_{50} of 2.93 mmHg, a pO_2 necessary to obtain 50% O_2 saturation of Mb. Under these conditions, O_2 most likely saturates cytochrome oxidase [69]. Raising the cellular pO_2 cannot directly enhance the cytochrome oxidase kinetics, which mirrors the VO_2. In fact, human gastrocnemius muscle studies show a decrease, not an increase, in pO_2 during exercise, in agreement with the cryosection analysis of canine gracilis muscle [59, 70]. In contrast, seal muscle during apnea shows Mb desaturation as blood flow and respiration decrease [61]. The intracellular O_2 supply then decreases when respiration either increases or decreases. As such, the O_2 supply alone cannot govern respiration rate in a simple fashion.

Although the fall in cellular pO_2 increases the O_2 gradient from capillary to cell, O_2 delivery from the cytosol to mitochondria must increase to accommodate increasing O_2 demand. A switch from free O_2 diffusion to Mb-facilitated diffusion to deliver O_2 to mitochondria could potentially compensate the falling pO_2 during muscle contraction. However, tissues with inhibited Mb function or without Mb do not exhibit any apparent handicap in respiration [71–73].

Indeed, during muscle contraction MbO_2 desaturates rapidly to a steady state with a time constant of 30 s. The desaturated steady-state levels rise with workload, but the kinetics time constant remains unchanged. Such rapid Mb desaturation implies that respiration does not operate necessarily in a kinetically slow way, consistent with an Eadie-Hofstee kinetic analysis of the intracellular VO_2 (based on Mb desaturation kinetics) versus ADP, which reveals no ADP-dependent regulation at the onset of muscle contraction [10].

8.14 Summary

^{1}H NMR can follow the distinct Mb and Hb signals and provide a basis for deconvoluting these signals in NIRS spectra [74]. Because the near-infrared method has a higher signal sensitivity than magnetic resonance, it can detect changes in vascular and intracellular O_2 more sensitively and on a faster timescale. However, NIRS has a much shallower penetration depth than NMR. A combined NMR and NIRS approach would help resolve many questions about the coordination of vascular and metabolic controls to meet the rapidly changing energy demand during muscle contraction, which in turn would provide invaluable insight into pressing questions about respiratory control under in-vivo conditions [74].

Acknowledgments We would like to acknowledge funding support from NIH GM 58688, and the Japan Society for the Promotion of Science, Bilateral Programs (7301001471) for help in writing the chapter and in obtaining some of the data described herein.

Problems

8.1. Marine mammals have a much higher concentration of Mb in their skeletal muscle than terrestrial mammals. What is the typical Mb concentration range in marine and terrestrial mammals? Why would an increase in Mb concentration confer a more prominent role for Mb facilitated O_2 diffusion? Discuss the role of Mb-facilitated O_2 diffusion in terms of O_2 diffusivity, Mb diffusivity, O_2 concentration, and Mb concentration in marine and terrestrial mammal myocytes. Assume identical O_2 and Mb diffusivity in all muscle cells.

8.2. A ^{1}H NMR experimental measurement of the deoxy-Mb signals in human gastrocnemius muscle shows no deoxy-Mb at rest. When the subject starts plantar flexion exercise at the rate of 70 contractions per minute, the deoxy-Mb signal appears and rises exponentially, consistent with the monoexponential relationship, $y = c - c^* \exp(-x/\tau)$ (y = Mb concentration (mM), x = time (sec), τ = time constant (sec), and c = arbitrary constant (mM)). During the first 2 min of exercise, venous pO_2 does not change, suggesting that almost all of the initial increase in muscle O_2 consumption arises from O_2 released from intracellular Mb. If $\tau = 30$ s, what is the mitochondrial respiration rate at the onset of plantar flexion exercise? What does the mitochondrial respiration at the onset of muscle contraction imply about the contribution of oxidative phosphorylation? Use a gastrocnemius Mb concentration estimate of 0.4 mM.

Further Reading

Ferrari M, Muthalib M, Quaresima V (2011) The use of near-infrared spectroscopy in understanding skeletal muscle physiology: recent developments. Philos Trans R Soc A 369:1–14

Gros G, Wittenberg BA, Jue T (2010) Myogloblin's old and new clothes: from molecular structure to function in living cells. J Exp Biol 213:2713–2725

References

1. From AH, Zimmer SD, Michurski SP, Mohanakrishnan P, Ulstad VK, Thoma WJ, Uğurbil K (1990) Regulation of the oxidative phosphorylation rate in the intact cell. Biochemistry 29(15):3731–3743
2. Chance B, Leigh JS, Kent J, McCully K, Niokam S, Clark BJ, Maris JM, Graham T (1986) Multiple controls of oxidative metabolism in living tissues as studied by phosphorus magnetic resonance. Proc Natl Acad Sci U S A 83:9458–9462
3. Erecinska M, Wilson DF (1982) Regulation of cellular metabolism. J Membr Biol 70:1–14
4. Heineman FW, Balaban RS (1990) Control of mitochondrial respiration in the heart in vivo. Annu Rev Physiol 52:523–542
5. Tschakovsy ME, Hughson RL (1999) Interaction of factors determining oxygen uptake at the onset of exercise. J Appl Physiol 86(4):1101–1113
6. Balaban RS, Kantor HL, Katz LA, Briggs RW (1986) Relation between work and phosphate metabolites in the in vivo paced mammalian heart. Science 232:1121–1123

7. Heineman FW, Kupriyanov VV, Marshall R, Fralix TA, Balaban RS (1992) Myocardial oxygenation in the isolated working rabbit heart as a function of work. Am J Physiol 262:H255–H267
8. Kreutzer U, Merkhamer Y, Tran TK, Jue T (1998) Role of oxygen in limiting respiration in the in situ myocardium. J Mol Cell Cardiol 30(12):2651–2655
9. Barstow TJ, Buchthal SD, Zanconato S, Cooper DM (1994) Changes in potential controllers of human skeletal muscle respiration during incremental calf exercise. J Appl Physiol 77:2169–2176
10. Chung Y, Mole PA, Sailasuta N, Tran TK, Hurd R, Jue T (2005) Control of respiration and bioenergetics during muscle contraction. Am J Physiol Cell Physiol 288(3):C730–C738
11. Wagner PD (1995) Muscle O_2. Med Sci Sports Exerc 27(1):47–53
12. Whipp BJ, Wasserman K (1972) Oxygen uptake kinetics for various intensities of constant-load work. J Appl Physiol 33(3):351–356
13. Albanese RA (1971) Use of membrane-covered oxygen cathodes in tissue. J Theor Biol 33:91–103
14. Whalen WJ, Nair P, Gainfield RA (1973) Measurements of oxygen tension in tissues with a micro oxygen electrode. Microvasc Res 5:254–262
15. Albanese RA (1973) On microelectrode distortion of tissue oxygen tensions. J Theor Biol 38:143–154
16. Whalen WJ, Riley J, Nair P (1967) A microelectrode for measuring intracellular pO_2. J Appl Physiol 23:798–801
17. Whalen WJ (1980) A hypodermic needle PO_2 electrode. J Appl Physiol 48:186–187
18. Lubbers DW (1969) The meaning of the tissue oxygen distribution curve and its measurement by means of Pt electrodes. Prog Respir Res 3:112–123
19. Inch WR (1958) Problems associated with the use of the exposed platinium electrode for measuring oxygen tension in vivo. Can J Biochem Physiol 36(10):1009–1021
20. Duling BR, Berne RM (1970) Longitudinal gradients in periarteriolar oxygen tension: a possible mechanism for the participation of oxygen in local regulation of blood flow. Circ Res 27:669–678
21. Antonini E, Brunori M (1971) Hemoglobin and myoglobin in their reactions with ligands. Elsevier/North Holland, Amsterdam
22. Owens JW, Connor CJ (1988) Characteriztion of some low spin complexes of ferric hemeoctapeptide from cytochrome C. Inorg Chim Acta 151:107–116
23. Reynafarje B (1963) Simplified method for the determination of myoglobin. J Lab Clin Med 61:138–145
24. Masuda K, Truscott K, Lin PC, Kreutzer U, Chung Y, Sriram R, Jue T (2008) Determination of myoglobin concentration in blood-perfused tissue. Eur J Appl Physiol 104(1):41–48
25. van Staveren HJ, Moes CJM, Prahl SA, van Gemert MJC (1991) Light scattering in Intralipid-10% in the wavelength range of 400–1100 nm. Appl Opt 30:4507–4514
26. Mourant JR, Fuselier T, Boyer T, Johnson TM, Bigio J (1997) Predictions and measurements of scattering and absorption over broad wavelength ranges in tissue phantoms. Appl Opt 36:949–957
27. Pogue BW, Patterson MS (2006) Review of tissue simulating phantoms for optical spectroscopy, imaging and dosimetry. J Biomed Opt 11:041102–041116
28. Strangman G, Boas DA, Sutton JP (2002) Non-invasive neuroimaging using near-infrared light. Biol Psychiatry 52:679–693
29. McCully K, Hamaoka T (2000) Near-infrared spectroscopy: what can it tell us about oxygen saturation in skeletal muscle? Exerc Sport Sci Rev 3(123):127
30. Boushel R, Piantoadosi CA (2000) Near-infrared spectroscopy for monitoring muscle oxygenation. Acta Physiol Scand 168:615–622
31. Ferrari M, Mottola L, Quaresima V (2004) Principles, techniques, of near-infrared spectroscopy. Can J Appl Physiol 29:463–487
32. Yamamoto K, Niwayama M, Shiga T, Lin L, Kudo N, Takahashi M (1998) Accurate NIRS measurement of muscle oxygenation by correcting the influence of a subcutaneous fat layer. Proc SPIE 3194:166–173
33. Niwayama M, Lin L, Shao J, Kudo N, Yamamoto K (2000) Quantitative measurement of muscle hemoglobin oxygenation using near-infrared spectroscopy with correction for the influence of a subcutaneous fat layer. Rev Sci Instrum 71:4571–4575
34. Costes F, Barthelemy JC, Feasson L, Busso T, Geyssant A, Denis C (1996) Comparison of muscle near-infrared spectroscopy and femoral blood gases during steady-state exercise in humans. J Appl Physiol 80:1345–1350
35. Theorell H (1938) Kristallinisches myoglobin. Biochem Z 268:73–82
36. Wilson JR, Mancini DM, McCully K, Feraro N, Lanoce V, Chance B (1989) Noninvasive detection of skeletal muscle underperfusion with near-infrared spectroscopy in patients with heart failure. Circulation 80:1668–1674
37. Harper AJ, Ferreira LF, Lutjemeier BJ, Townsend DK, Barstow TJ (2006) Human femoral artery and estimated muscle capillary blood flow kinetics following the onset of exercise. Exp Physiol 91:661–671
38. Kindig CA, Richardson TE, Poole DC (2002) Skeletal muscle capillary hemodynamics from rest to contractions: implications for oxygen transfer. J Appl Physiol 92:2513–2520

39. Bank W, Chance B (1994) An oxidative defect in metabolic myopathies: diagnosis by noninvasive tissue oximetry. Ann Neurol 36:830–837
40. Marcinek DJ, Amara CE, Matz K, Conley KE, Schenkman KA (2007) Wavelength shift analysis: a simple method to determine the contribution of hemoglobin and myoglobin to in vivo optica spectra. Appl Spectrosc 61:665–669
41. Binzoni T, Cooper CE, Wittekind AL, Beneke R, Elwell CE, Van De Ville D, Leung TS (2010) A new method to measure local oxygen consumption in human skeletal muscle during dynamic exercise using near-infrared spectroscopy. Physiol Meas 31:1257–1269
42. Jue T, Anderson S (1990) ^{1}H observation of tissue myoglobin: an indicator of intracellular oxygenation in vivo. Magn Reson Med 13:524–528
43. Dayhoff MO, Eck RV (1968) Atlas of protein sequence and structure. National Biomedical Research Foundation, Silver Spring
44. Kendrew JC (1963) Myoglobin and the structure of proteins. Science 139:1259–1266
45. Weissbluth M (1974) Cooperativity and electronic properties. Mol Biol Biochem Biophys 15:1–175
46. Horrocks JD (1973) Analysis of isotropic shifts. In: La Mar GN, Horrocks JD, Holm RH (eds) NMR of paramagnetic molecules. Academic, New York, pp 128–175
47. Jesson JP (1973) In: La Mar GN, Horrocks JD, Holm RH (eds) The paramagnetic shift in NMR of paramagnetic molecules. Academic, New York, pp 1–51
48. La Mar GN, Horrocks WD Jr, Holm RH (eds) (1973) NMR of paramagnetic molecules. Academic, New York
49. La Mar GN (1979) Biological applications of magnetic resonance. Academic, New York
50. Goff H, La Mar GN (1977) High-spin ferrous porphyrin complexes as models for deoxymyoglobin and deoxyhemoglobin: a proton nuclear magnetic resonance study. J Am Chem Soc 99:6599–6606
51. La Mar GN, Budd DL, Goff H (1977) Assignment of proximal histidine proton NMR peaks in myoglobin and hemoglobin. Biochem Biophys Res Commun 77:104–110
52. La Mar GN, Nagai K, Jue T, Budd DL, Gersonde K, Sick H, Kagimoto T, Hayashi A, Taketa F (1980) Assignment of proximal histidyl imidazole exchangeable proton NMR resonances to individual subunits in hemoglobins A, Boston, Iwate and Milwaukee. Biochem Biophys Res Commun 96:1172–1177
53. Ho C, Russu I (1981) Proton nuclear magnetic resonance investigation of hemoglobins. Methods Enzymol 76:275–312
54. Kreutzer U, Chung Y, Butler D, Jue T (1993) ^{1}H-NMR characterization of the human myocardium myoglobin and erythrocyte hemoglobin signals. Biochim Biophys Acta 1161:33–37
55. Jue T, La Mar GN, Han K, Yamamoto Y (1984) NMR study of the exchange rates of allosterically responsive labile protons in the heme pockets of hemoglobin A. Biophys J 46:117–120
56. Patel DL, Kampa L, Shulman RG, Yamane T, Wyluda BJ (1970) Proton nuclear magnetic resonance studies of myoglobin in H_2O. Proc Natl Acad Sci U S A 67:1109–1115
57. Lindstrom TR, Ho C (1972) Functional nonequivalence of α and β hemes in human adult hemoglobin. Proc Natl Acad Sci U S A 69(7):1707–1710
58. Kreutzer U, Wang DS, Jue T (1992) Observing the ^{1}H NMR signal of the myoglobin Val-E11 in myocardium: an index of cellular oxygenation. Proc Natl Acad Sci U S A 89:4731–4733
59. Mole PA, Chung Y, Tran TK, Sailasuta N, Hurd R, Jue T (1999) Myoglobin desaturation with exercise intensity in human gastrocnemius muscle. Am J Physiol 277(1 Pt 2):R173–R180
60. Ponganis PJ, Kreutzer U, Sailasuta N, Knower T, Hurd R, Jue T (2002) Detection of myoglobin desaturation in *Mirounga angustirostris* during apnea. Am J Physiol Regul Integr Comp Physiol 282:R267–R272
61. Ponganis PJ, Kreutzer U, Stockard TK, Lin PC, Sailasuta N, Tran TK, Hurd R, Jue T (2008) Blood flow and metabolic regulation in seal muscle during apnea. J Exp Biol 211(Pt 20):3323–3332
62. Masuda K, Takakura H, Furuichi Y, Iwase S, Jue T (2010) NIRS measurement of O_2 dynamics in contracting blood and buffer perfused hindlimb muscle. Adv Exp Med Biol 662:323–328
63. Mancini DM, Wilson JR, Bolinger L, Li H, Kendrick K, Chance B, Leigh JS (1994) In vivo magnetic resonance spectroscopy measurement of deoxymyoglobin during exercise in patients with heart failure. Circulation 90:500–508
64. Ferrari M, Muthalib M, Quaresima V (2011) The use of near-infrared spectroscopy in understanding skeletal muscle physiology: recent developments. Philos Trans R Soc Lond 369:1–14
65. Nioka S, Wang DJ, Im J, Hamaoka T, Wang ZJ, Leigh JS, Chance B (2009) Simulation of Mb/Hb in NIRS and oxygen gradient in the human and canine skeletal muscle using H-NMR and NIRS. Adv Exp Med Biol 578:223–228
66. Hoofd L, Colier W, Oeseburg B (2009) A modeling investigation to the possible role of myoglobin in human muscle in near infrared spectroscopy (NIRS) measurements. Adv Exp Med Biol 530:637–643
67. Jue T (1994) Measuring tissue oxygenation with the ^{1}H NMR signals of myoglobin. In: Gilles R (ed) NMR in physiology and biomedicine. Academic, New York, pp 199–207

68. Kreutzer U, Jue T (1991) ^{1}H-nuclear magnetic resonance deoxymyoglobin signal as indicator of intracellular oxygenation in myocardium. Am J Physiol Cell Physiol 261(6 pt 2):H2091–H2097
69. Chance B (1989) Metabolic heterogeneities in rapidly metabolizing tissue. J Appl Cardiol 4:207–221
70. Gayeski TEJ, Honig CR (1988) Intracellular Po_2 in long axis of individual fiber in working dog gracilis muscle. Am J Physiol 254:H1179–H1185
71. Flogel U, Merx MW, Godecke A, Decking UKM, Schrader J (2001) Myoglobin: a scavenger of bioactive NO. Proc Natl Acad Sci U S A 98(2):735–740
72. Garry DJ, Ordway GA, Lorenz JN, Radford NB, Chin ER, Grange RW, Bassel-Duby R, Williams RS (1998) Mice without myoglobin. Nature 395:905–908
73. Glabe A, Chung Y, Xu D, Jue T (1998) Carbon monoxide inhibition of regulatory pathways in myocardium. Am J Physiol 274:H2143–H2151
74. Tran TK, Sailasuta N, Kreutzer U, Hurd R, Chung Y, Mole P, Kuno S, Jue T (1999) Comparative analysis of NMR and NIRS measurements of intracellular pO_2 in human skeletal muscle. Am J Physiol 276:R1682–R1690

Problem Solutions

9

Chapter 1

1.1. How can weak photocurrent of an Si photodiode be converted to a voltage signal on continuous-wave NIRS or spatially resolved NIRS?

Answer

A current-to-voltage converter using an operational amplifier is generally used to convert a weak current signal I. A photodiode is connected to the input of an operational amplifier. When a feedback register of the circuit is R_f, the output voltage is simply $R_f I$. The design of the converter requires a lot of trial and error in order to reduce noise and bias. For the details, refer to the articles listed under Further Reading.

Chapter 2

2.1. A random number used on Monte Carlo simulation should be long period and of almost uniform distribution. How can this random number be generated?

Answer

A Mersenne Twister (MT) is a pseudorandom number–generating algorithm developed by Matsumoto and Nishimura in 1997. It has the following features: (1) a far longer period and a far higher order of equidistribution than any other implemented generator, (2) fast generation, and (3) efficient use of memory. For example, the implemented C-code mt19937.c consumes only 624 words of working area and provides a random number that has a period of $2^{19937} - 1$ and a 623-dimensional equidistribution property. Please refer to the papers cited in Further Reading for details.

Chapter 3

3.1. Assume that the concentration of hemoglobin is changed from 0.1 to 0.11 mM and the oxygen saturation of the blood is changed from 65% to 70% in the activated region of the brain. The extinction coefficients of oxygenated hemoglobin and deoxygenated hemoglobin at 780-nm

T. Jue and K. Masuda (eds.), *Application of Near Infrared Spectroscopy in Biomedicine*, Handbook of Modern Biophysics 4, DOI 10.1007/978-1-4614-6252-1_9,

wavelength are 0.16 and 0.25 mM^{-1} mm^{-1}, respectively. The partial optical pathlength in the activated region for a probe pair is 5 mm.

(a) Find the change in absorption coefficient at 780-nm wavelength in the activated region.

Answer

The concentration change in the oxygenated hemoglobin $\Delta c_{\text{oxy-Hb}}$ and deoxygenated hemoglobin $\Delta c_{\text{dexoy-Hb}}$ can be calculated as follows:

$$\Delta c_{\text{oxy-Hb}} = 0.11\ [\text{mM}] \times 70\,\% - 0.1[\text{mM}] \times 65\% = 0.012\ [\text{mM}],$$

$$\Delta c_{\text{deoxy-Hb}} = 0.11\ [\text{mM}] \times 30\,\% - 0.1[\text{mM}] \times 35\,\% = -0.002\ [\text{mM}].$$

The change in absorption coefficient is

$$\begin{aligned}\Delta\mu_a &= \varepsilon_{\text{oxy-Hb}} \cdot \Delta c_{\text{oxy-Hb}} + \varepsilon_{\text{deoxy-Hb}} \cdot \Delta c_{\text{deoxy-Hb}} \\ &= 0.16\ [\text{mM}^{-1}\text{mm}^{-1}] \times 0.012\ [\text{mM}] + 0.25\ [\text{mM}^{-1}\text{mm}^{-1}] \times -0.002\ [\text{mM}] \\ &= 1.42 \times 10^{-3}[\text{mm}^{-1}].\end{aligned}$$

(b) Find the change in the optical density (NIRS signal) at 780-nm wavelength caused by the absorption change in the activated region:

$$\Delta\text{OD} = \Delta\mu_a \cdot <L_{\text{act}}>= 1.42 \times 10^{-3}[\text{mm}^{-1}] \times 5\ [\text{mm}] = 7.1 \times 10^{-3}.$$

3.2. Derive the equations that calculate the concentration change in oxygenated and deoxygenated hemoglobins from change in optical density (NIRS signal) at two wavelengths, λ_1 and λ_2. The extinction coefficient of oxygenated hemoglobin and deoxygenated hemoglobin is $\varepsilon_{\text{oxy-Hb}}$ and $\varepsilon_{\text{deoxy-Hb}}$, respectively. (Assume that the wavelength dependence of the partial optical pathlength in the activated region $<L_{\text{act}}>$ can be ignored.)

Answer

The relationship between the NIRS signal at two wavelengths and change in oxygenated and deoxygenated hemoglobins is

$$\begin{aligned}\Delta\text{OD}(\lambda_1) &= (\varepsilon_{\text{oxy-Hb}}(\lambda_1) \cdot \Delta c_{\text{oxy-Hb}} + \varepsilon_{\text{deoxy-Hb}}(\lambda_1) \cdot \Delta c_{deoxy-Hb}) \cdot <L_{\text{act}}>, \\ \Delta\text{OD}(\lambda_2) &= (\varepsilon_{\text{oxy-Hb}}(\lambda_2) \cdot \Delta c_{\text{oxy-Hb}} + \varepsilon_{\text{deoxy-Hb}}(\lambda_2) \cdot \Delta c_{\text{deoxy-Hb}}) \cdot <L_{\text{act}}>.\end{aligned}$$

The equations, which calculate the concentration change in oxygenated and deoxygenated hemoglobins, can be derived by solving the above simultaneous equations:

$$\Delta c_{\text{oxy-Hb}} = \frac{\Delta\text{OD}(\lambda_2)\varepsilon_{\text{deoxy-Hb}}(\lambda_1) - \Delta\text{OD}(\lambda_1)\varepsilon_{\text{deoxy-Hb}}(\lambda_2)}{(\varepsilon_{\text{oxy-Hb}}(\lambda_2)\varepsilon_{\text{deoxy}}(\lambda_1) - \varepsilon_{\text{oxy-Hb}}(\lambda_1)\varepsilon_{\text{deoxy-Hb}}(\lambda_2))\ \langle L_{\text{act}}\rangle}.$$

$$\Delta c_{\text{deoxy-Hb}} = \frac{\Delta\text{OD}(\lambda_1)\varepsilon_{\text{oxy-Hb}}(\lambda_2) - \Delta\text{OD}(\lambda_2)\varepsilon_{\text{oxy-Hb}}(\lambda_1)}{(\varepsilon_{\text{oxy-Hb}}(\lambda_2)\varepsilon_{\text{deoxy-Hb}}(\lambda_1) - \varepsilon_{\text{oxy-Hb}}(\lambda_1)\varepsilon_{\text{deoxy-Hb}}(\lambda_2))\ \langle L_{\text{act}}\rangle}.$$

3.3. Draw polar plots of the probability distribution of deflection angle $p(\theta)$ described by the Henyey-Greenstein phase function for $g = 0.1$, $g = 0.5$, and $g = 0.9$.

Answer

The probability distribution of the deflection angle can be calculated by the following equation:

$$p(\theta) = \frac{1 - g^2}{4\pi(1 + g^2 - 2g\cos\theta)^{\frac{3}{2}}},$$

3.4. A pencil beam of a short pulse is incident onto tissues and diffusely reflected light is detected at 20 mm from the incident point. Analyze light propagation in the tissues by analytical solution of the diffusion equation described in [26]. The optical properties of the tissues: (1) $\mu_s = 10\ \text{mm}^{-1}$, $g = 0.9$, $\mu_a = 0.01\ \text{mm}^{-1}$. (2) $\mu_s = 10\ \text{mm}^{-1}$, $g = 0.85$, $\mu_a = 0.01\ \text{mm}^{-1}$. (3) $\mu_s = 5\ \text{mm}^{-1}$, $g - 0.8, \mu_a - 0.02\ \text{mm}^{-1}$. Although the diffusion coefficient is defined as $\kappa = 1/3\{(\mu'_s + \mu_a)\}$ in [26], $\kappa = 1/(3\mu'_s)$ can be used for the calculations. The speed of light in the medium is 0.2 mm/ps, and refractive index mismatch at the tissue boundary can be ignored.

(a) Determine the transport scattering coefficient of each tissue.

Answer

The transport scattering coefficient μ'_a is calculated using the scattering coefficient μ_s and anisotropic factor g by the following equation:

$$\mu'_a = (1 - g)\mu_s.$$

1. $\mu'_s = (1 - 0.9) \times 10\ [\text{mm}^{-1}] = 1.0\ [\text{mm}^{-1}]$,
2. $\mu'_s = (1 - 0.85) \times 10\ [\text{mm}^{-1}] = 1.5\ [\text{mm}^{-1}]$,
3. $\mu'_s = (1 - 0.8) \times 5\ [\text{mm}^{-1}] = 1.0\ [\text{mm}^{-1}]$.

(b) Determine the depth of the isotropic point source created by the incident beam.

Answer

The depth of the isotropic point source z_0 is the reciprocal of the transport scattering coefficient:

$$z_0 = 1/\mu'_s.$$

1. $z_0 = 1/\mu'_s = 1/1.0\ [\text{mm}^{-1}] = 1.0\ [\text{mm}]$,
2. $z_0 = 1/\mu'_s = 1/1.5\ [\text{mm}^{-1}] = 0.667\ [\text{mm}]$,
3. $z_0 = 1/\mu'_s = 1/1.0\ [\text{mm}^{-1}] = 1.0\ [\text{mm}]$.

(c) Draw the temporal distribution of the reflectance.

Answer

The temporal distribution of the reflectance is given by

$$R(\rho, t) = (4\pi\kappa c)^{-3/2} z_0 t^{-5/2} \exp(\mu_a ct) \exp\left(-\frac{\rho^2 + {z_0}^2}{4\kappa ct}\right),$$

where ρ is the distance between the detection and incident points, c is the speed of light in the tissue, and t is time.

Chapter 4

4.1. How does NIRS help us better understand the relationship between vasculature and tissue function?

Answer

See Further Reading.

Chapter 5

5.1. How would you quantify muscle NIR signals?

5.2. List the various muscle NIR indicators. Which indicator reflects muscle oxidative function? How?

Answer

See Further Reading.

Chapter 6

6.1. NIRS measurement of Mb saturation (S_{MbO2}) at different tensions during muscle contraction shows the following:

Tension (%)	S_{MbO2} (%)
50	70
75	59
10	49

What is the corresponding change in PO_2, given an Mb P50 of 2.37 at 37°C? What is the change in O_2 gradient, given a resting PO_2 of 10 mmHg? Plot out the curves. Does the change in S_{MbO2}, PO_2, and O_2 gradient show a linear relationship? What is the physiological implication in interpreting the S_{MbO2} data with respect to O_2 gradient?

Answer

See Takakura H, Masuda K, Hashimoto T, Iwase S, Jue T (2010) Quantification of myoglobin deoxygenation and intracellular partial pressure of O_2 during muscle contraction during haemoglobin-free medium perfusion. Exp Physiol 95:630–640

Chapter 7

7.1. What is the ratio of Mb to Hb in a gram of the locomotory muscle of a seal versus that of a dog? (see [41–45]).

Answers

1. To solve this problem, we need to first determine the μmol content of Mb in 1 g of locomotory muscle in the seal and the dog:
   (a) Using seal Mb concentration and its molecular weight, the micromole content is easily calculated.

      Seal [Mb] is 37 mg/g of the epaxial muscle (from [44]).

Note: Different [Mb] or [Hb] concentrations or capillary characteristics may be found in different studies or for different muscles. However, although the calculations may change, the trends demonstrated in this problem will be the same.
The gram content of Mb in 1 g of muscle is converted to micromoles as follows:

$$\text{Mb} = 37\ \text{mg} = 0.037\ \text{g},$$
$$\text{Mb molecular weight} = 17{,}000\ \text{g/mol},$$
$$\text{Mb} = 0.037\text{g}/17{,}000\ \text{g/mol} = 2.18 \times 10^{-6}\text{mol},$$
$$2.18 \times 10^{-6}\text{mol} = 2.18\ \mu\text{mol},$$
$$\text{Seal Mb} = 2.18\ \mu\text{mol}.$$

(b) The same calculation is done for the dog Mb:
Dog [Mb] is 1.5 mg/g in the gastrocnemius (provided from [44]).
Again Mb is converted to micromoles:

$$\text{Mb} = 1.5\ \text{mg} = 0.0015\ \text{g},$$
$$\text{Mb molecular weight} = 17{,}000\ \text{g/mol},$$
$$\text{Mb} = 0.0015\ \text{g}/17{,}000\ \text{g/mol} = 8.82 \times 10^{-8}\text{mol},$$
$$\text{Mb} = 8.82 \times 10^{-8}\text{mol} = 0.0882\ \mu\text{mol}.$$
$$\text{Dog Mb} = 0.0882\ \mu\text{mol}.$$

2. Next, the μmol content of Hb in 1 g of locomotory muscle must be determined. To do this, the muscle capillary volume (V_c) must first be determined in a gram of muscle and then the content of Hb in that volume can be calculated. The values for both the seal and dog are provided in [41, 42] and the formula for Vc is in [43].
   (a) Using the formula and capillary data for harbor seals, the V_c in 1 g of locomotory muscle (longissimus dorsi) is calculated, which is then multiplied by the [Hb]:

$$V_c = (\text{capillary density in \# capillaries/mm}^2)\ (\text{capillary anisotropy coefficient})$$
$$(\pi/4)\ (\text{diameter mm}^2)/(\text{muscle density g/ml}),$$
$$\text{capillary density} = 1138/\text{mm}^2(\text{mean of several muscles}),$$
$$\text{capillary anisotropy coefficient} = 1.2,$$
$$\text{diameter} = 0.00449\ \text{mm},$$
$$\text{muscle density} = 1.06\ \text{g/ml},$$
$$\text{Seal } V_c = (1138/\text{mm}^2)\ (1.2)\ (0.785)\ (0.00449\ \text{mm})\ (0.00449\ \text{mm})/(1.06\ \text{g/ml}),$$
$$\text{Seal } V_c = 0.0204\ \text{ml}$$

Taking the concentration of [Hb] in a seal (provided from [42]), Hb content in the V_c per 1 g of muscle can be determined:

$$\text{Seal [Hb]} = 0.222\ \text{g/ml},$$
$$\text{Hb in } V_c = (0.0204\ \text{ml})\ (0.222\ \text{g/ml}),$$
$$\text{Hb in } V_c = 0.0045\ \text{g},$$
$$\text{Hb molecular weight} = 64{,}500\ \text{g/mol},$$
$$\text{Hb in } V_c = 0.0045\ \text{g}/64{,}500\ \text{g/mol},$$
$$\text{Hb in } V_c = 6.97 \times 10^{-8}\text{mol},$$
$$\text{Seal Hb in 1 gm muscle} = 0.070\ \mu\text{mol}.$$

(b) Using the formula and capillary data for dogs, the V_c in 1 g of locomotory muscle is calculated, which is then multiplied by the dog [Hb].
The calculation can be done for the dog using data from [41, 42]:

$$V_c = (\text{capillary density in \# capillaries/mm}^2)(\text{capillary anisotropy coefficient})$$
$$(\pi/4)\ (\text{diameter mm}^2)/(\text{muscle density g/ml}),$$
$$\text{capillary density} = 1{,}617/\text{mm}^2(\text{mean of several muscles}),$$
$$\text{capillary anisotropy coefficient} = 1.23,$$
$$\text{diameter} = 0.00416\ \text{mm},$$
$$\text{muscle density} = 1.06\ \text{g/ml}.$$
$$\text{Dog } V_c = (1{,}617/\text{mm}^2)\ (1.23)\ (0.785)(0.00416\ \text{mm})\ (0.00416\ \text{mm})/\ (1.06\ \text{g/ml}),$$
$$\text{Dog } V_c = 0.0255\ \text{mL}.$$

Taking the concentration of [Hb] in a dog, provided from the citation below, Hb content in the V_c in 1 g of muscle can be determined (see [42]):

$$\text{Dog [Hb]} = 0.188\ \text{g/ml},$$
$$\text{Hb in } V_c = (0.0255\ \text{ml})\ (0.188\ \text{g/ml}),$$
$$\text{Hb in } V_c = 0.0048\ \text{g},$$
$$\text{Hb molecular weight} = 64{,}500\ \text{g/mol},$$
$$\text{Hb in } V_c = 0.0048\ \text{g}\ /\ (64{,}500\text{g/mol}),$$
$$\text{Hb in } V_c = 7.43 \times 10^{-8}\text{mol}.$$
$$\text{Dog Hb in 1 gm muscle} = 0.074\mu\text{mol}.$$

Ratio of Mb:Hb for the seal:
Seal Mb:Hb = 2.18 μmol/0.070 μmol,
Seal Mb:Hb = 31.1:1.
Ratio of Mb:Hb for the dog:
Dog Mb:Hb = 0.0882 μmol/0.074 μmol,
Dog Mb:Hb = 1.19:1.
Ratios of Mb: Hb in Muscle:
Seal: 31.1:1,
Dog: 1.19:1.

Seal Mb: Dog Mb:

26.1:1.

These calculations, based on Mb concentration, Hb concentration, and muscle capillary density, reveal that the ratio of Mb to Hb in seal muscle is 26 times greater than in dog muscle. This is primarily a reflection of the 25-fold greater Mb concentration in seal muscle. Although Hb concentration is lower in the dog, capillary density is slightly greater in the dog than in the seal. These ratios are representative of the resting state. During exercise in dogs, increased muscle blood flow and muscle capillary recruitment would make the dog's Mb-to-Hb ratio even lower. In contrast, in the diving seal, even if muscle blood flow is only partially reduced, the Mb-to-Hb ratio will be increased. Thus, although it is difficult to quantify the contribution of Hb to the NIRS signal in muscle, the high Mb concentration and the cardiovascular responses in the diving seal minimize the potential contribution of Hb to the NIRS signal in muscle.

Chapter 8

8.1. Marine mammals have a much higher of Mb in their skeletal muscle than terrestrial mammals. What is the typical Mb concentration range in marine and terrestrial mammals? Why would an increase in Mb concentration confer a more prominent role for Mb facilitated O_2 diffusion? Discuss the role of Mb-facilitated O_2 diffusion in terms of O_2 diffusivity, Mb diffusivity, O_2 concentration, and Mb concentration in marine and terrestrial mammal myocytes. Assume identical O_2 and Mb diffusivity in all muscle cells.

Answer

See Ponganis PJ, Kreutzer U, Sailasuta N, Knower T, Hurd R, Jue T (2002) Detection of myoglobin desaturation in *Mirounga angustirostris* during apnea. Am J Physiol Regul Integr Comp Physiol 282:R267–R272

8.2 A ^{1}H NMR experimental measurement of the deoxy-Mb signals in human gastrocnemius muscle shows no deoxy-Mb at rest. When the subject starts plantar flexion exercise at the rate of 70 contractions per minute, the deoxy-Mb signal appears and rises exponentially, consistent with the monoexponential relationship, $y = c - c^* \exp(-x/\tau)$ (y = Mb concentration (mM), x = time (s), τ = time constant (s), and c = arbitrary constant (mM)). During the first 2 min of exercise, venous pO_2 does not change, suggesting that almost all of the initial increase in muscle O_2 consumption arises from O_2 released from intracellular Mb. If $\tau = 30$ s, what is the mitochondrial respiration rate at the onset of plantar flexion exercise? What does the mitochondrial respiration at the onset of muscle contraction imply about the contribution of oxidative phosphorylation? Use a gastrocnemius Mb concentration estimate of 0.4 mM.

Answer

See Ponganis PJ, Kreutzer U, Stockard TK, Lin PC, Sailasuta N, Tran TK, Hurd R, Jue T (2008) Blood flow and metabolic regulation in seal muscle during apnea. J Exp Biol 211 (Pt 20): 3323–3332

Index

T. Jue and K. Masuda (eds.), *Application of Near Infrared Spectroscopy in Biomedicine*, Handbook of Modern Biophysics 4, DOI 10.1007/978-1-4614-6252-1,

N

O

MIX
Papier aus verantwortungsvollen Quellen
Paper from responsible sources
FSC® C105338

Printed by Books on Demand, Germany